The IDEALS *of* JOSEPH BEN-DAVID

The IDEALS *of* JOSEPH BEN-DAVID

The Scientist's Role and Centers of Learning Revisited

Liah Greenfeld, editor

Routledge
Taylor & Francis Group
LONDON AND NEW YORK

First published 2012 by Transaction Publishers

2 Park Square, Milton Park, Abingdon, Oxfordshire OX14 4RN
711 Third Avenue, New York, NY 10017

Routledge is an imprint of the Taylor & Francis Group, an informa business

First issued in paperback 2017

Library of Congress Catalog Number: 2011042806

Library of Congress Cataloging-in-Publication Data

The ideals of Joseph Ben-David : the scientist's role and centers of learning revisited / Liah Greenfeld, editor.
p. cm.
Includes index.
ISBN 978-1-4128-4293-8
1. Ben-David, Joseph. 2. Scientists—Israel—Biography. 3. Science and civilization. 4. Education, Higher. I. Greenfeld, Liah.
Q149.I8I34 2012
509.2—dc23

2011042806

ISBN 13: 978-1-4128-4293-8 (hbk)
ISBN 13: 978-1-138-51610-6 (pbk)

Contents

Preface and Acknowledgments

Joseph Ben-David died a long time ago in January 1986. An eminent sociologist of science, in fact a cofounder of the subdiscipline, he was only sixty-five years old. Few social scientists are remembered after they die and can no longer parlay their influence into goods of this world for colleagues and acquaintances. This was not his fate. Ben-David's work continued to be taught and referred to by scholars spread far and wide (in terms of both countries and disciplines); his students never forgot him; his books were republished and his essays appeared in new collections.

In the spring of 2007, I, as a student of Ben-David, was approached with a request to coorganize an international conference on the "efficiency" of modern universities, i.e., their ability to produce scientists and to encourage the production of science within their walls. Ben-David, among other things, wrote a seminal comparative study of universities, *Centers of Learning*, considered by many a contribution of utmost importance to the understanding of such "efficiency"; the proposed conference would thus be a homage to his memory. Though I have been writing on science on and off throughout my career, I would not define myself (and others do not see me) as a sociologist of science, and, as to "efficiency," I believe that it properly belongs on the conveyor line, rather than within institutions explicitly concerned with the mind and creativity. Yet, this was not a request I could deny. For weighty reasons of a personal, moral, and emotional character, I had no choice but to take this project on.

Joseph Ben-David was my only teacher; I considered no one else as one. We met in 1974, a sophomore year for me, in his class on "Science and Ideology," and clicked immediately: it was clear to both of us within the first fifteen minutes, when he asked his first question and I answered it after several attempts by others, that he must be my teacher and I must be his student. This is not how teacher–student relationships normally commence, and I consider it unusually auspicious when they start like that between my students and myself, now as a teacher. Ben-David taught me the importance of the teacher–student link and I always try to impart it to my students. This link is a very important element of the process of tradition in science and its progress. It also makes scholarship a life—not simply a career chosen among others (though obviously it is so chosen), but an essential part of one's identity—in which one's teachers and students become members of one's family. Organizing a conference in honor of my teacher quarter of a century after his death, in my view, was a continuation of such life and had the deep emotional significance of a communion. I envisioned it, among other things, as a celebration of the teacher–student link, an event of prime pedagogical importance. I made it a point to tell my current students about it and tried to involve in it students of Ben-David and those (my students and colleagues) who learned about Ben-David from me.

After I completed my doctoral dissertation in Israel, Ben-David, already very ill, sent me, as a mental package of sorts, to the University of Chicago, to study with his own teacher, Edward Shils. This not only created a direct link between the teacher of my teacher and me, but reinforced the links both between me and Ben-David and between Ben-David and Shils. That other great man took his responsibilities as a grand-teacher very seriously and for twelve years until his death watched over my intellectual development, making me conscious of being part of a living tradition. Ben-David died too early for me to connect my students to him in this personal and direct way, and yet, I felt I was able to transmit this consciousness to them. It was a high point of the conference for me when Professor Chikako Takeishi of Tokyo's Chuo University—my student—began her presentation by stressing how significant it was for her to speak about the teacher of her teacher.

Another part of the life of scholarship, as Ben-David taught it to me, was to regard the pursuit of objective knowledge as a civic responsibility and a moral duty. For Ben-David, all scholarship worthy of its name (and his commitment) was science, i.e., the pursuit of objective knowledge of empirical reality through the conscientious application of the method of conjectures and refutations. The end and the means were to guide the scholar's professional activity unswayed by his or her other interests, whether personal or political. While nobody could expect scientists to be value-free (if they were, they would be freaks of nature), science had to be value-free. The "autonomy of science," in this context, referred to its right to be value-free, which had to be zealously preserved. Scholarship oriented to any goal other than the accumulation of objective knowledge about empirical reality was science no longer and had no right to claim its authority. To use the authority of science for the achievement of nonscientific goals was a travesty, in effect, a sin. Life of scholarship was one of moral dedication, and nothing less than the fate of the liberal democratic society depended on it, because science was the embodiment of the cardinal virtue of this society: the virtue of civility. Ben-David had never said this in so many words; he taught me by example: he embodied the principles of value-free, autonomous science in his research, in his work as a teacher, in his life. To celebrate—and reaffirm—these principles, no longer fashionable in the free countries in which freedom is seen as a natural endowment and not a precious achievement to be cherished and safeguarded, seemed to me the most meaningful way to honor his memory.

The conference, entitled "The Role of the University in the 21st Century," took place in Switzerland in July 2009. The funds for it were raised by my coorganizer and the originator of the idea for the project, a retired university administrator from the Zurich Technological University, Marcel Herbst. While it was proven beyond reasonable doubt that most universities around the world, sadly, have become woefully "inefficient," it also became clear that the ideals and ideas of my teacher continue to inform and inspire scholars in many countries and disciplines. I am grateful to Mr. Herbst for coming up with the idea of the conference and the means to bring most of the authors in this collection face to face. I am grateful to the authors—participants in the conference—for persevering through a much-too-long process of preparing this collection for publication; to my friend and colleague Brendan Dooley, who joined us at the very last moment, for contributing a thought-provoking piece on the distinctiveness of science; to the dear, brave Miriam Ben-David for allowing me to include in lieu of an epilogue the photographs of her late husband she annotated for the presentation at the conference. And I am very, very grateful to Irving L. Horowitz, Mary Curtis, and the staff of Transaction Publishers for staying with the project, despite some quite unusual (though not entirely unexpected, given the state of the Western academia) trying circumstances, and for including this volume in the list of their fiftieth anniversary year. Finally, I must thank Jorge Montesdeoca Perez, a true *hidalgo,* if ever there was one, for volunteering to compile the index for this book.

This volume is dedicated to the children, grandchildren, and great-grandchildren of Joseph Ben-David, and to that other line of his descendants, our students.

September 2011
Belmont, Massachusetts

Introduction

This volume brings together some of the proceedings of the conference on "The Role of the University in the 21st Century." The conference, among other things, re-examined the ideas of Joseph Ben-David on the nature of creativity and progress in science, and the ability of the research universities, since the nineteenth century the undisputed centers of secular learning, to combine scientific training with the no less important task allotted to them by modern society of moral (civic or character) education of the young. In light of the widespread recognition of the transformation of the universities in Western liberal democracies from knowledge-seeking institutions, with the central purpose of creating a congenial environment for free generation of ideas, into bureaucracies driven by the interest of self-perpetuation and thus centrally preoccupied with generating revenue,[1] the reduced competitiveness of Western scientific establishments vis-à-vis those of the rising Asian powers, and the general loss of self-confidence among Western nations, such a re-examination is timely and has significance beyond the confines of a discourse on science and higher education. The nine papers written for the conference are supplemented with two more: Ben-David's own early statement on the former question (the nature of scientific creativity and progress), published in Hebrew in 1964, and Brendan Dooley's essay on science vs. astrology, which open the collection. The first six essays following Ben-David's (Dooley, Panchenko, Gellert, Coccia, Takeishi, and Ezrahi) address the nature and conditions of creativity in science in general; the next two (Richard and Greenfeld) focus on the social sciences, while the following two papers (Wood and Ben-Chaim) have education as their focus. The authors of the essays included here come from several countries (France, Germany, Israel, Italy, Japan, Russia, and the United States) and disciplines (anthropology, classics, early modern Italian history, economics, intellectual history, history and philosophy of science, political science, and sociology), demonstrating the lasting relevance of Ben-David's ideas and broad significance of his central subjects.

Science emerged as one of the most authoritative social institutions when the new, secular form of consciousness, nationalism, refocused the attention of humanity under its sway on this world, for the first time in the history of the West, at least, placing empirical reality on a par with, and soon after above, transcendental realms. Now seen as ultimately meaningful in its own right and generating its own laws, empirical reality—whether physical or social—had to be carefully studied, and the epistemic system dedicated to its study, science, in modern society acquired the dignity of theology. Because the institutionalization of science coincided with the emergence of nationalism, which happened in England in the sixteenth century, and because England turned to science explicitly for reasons of national aggrandizement,[2] science from the outset became a central (maybe even the central) area of competition for national prestige. In the eyes of the lay public, in particular, it came to reflect the intellectual giftedness of a people, its ability to contribute to the common stores of human knowledge, and therefore its overall importance for humanity. Superiority in science, since early seventeenth century in England, has been considered the very expression of the superiority of the national character, the measure of a nation's smarts.

This has been so in all scientifically competitive nations—first in England, then in France, Germany, Russia, the United States, Japan, now in China and India—being, obviously, most pronounced when the nation clearly led the race, and waning when it was less competitive; so far, only the moral standing of a nation (its dedication to the modern values of freedom and equality) and the strength of its economy—both, in fact, in the popular imagination related to science—could compete with science as the gauge of the overall national worth. At the outset, this was not different in the young nation of Israel. When Joseph Ben-David wrote his paper "Science in a Small Country," Israel was sixteen years old and the oldest university in it—the Hebrew University of Jerusalem, where scientific research concentrated—was, like Ben-David himself, about forty. Unrestrained optimism regarding the superiority of the Israeli scientific establishment, apparently, reigned both in the academic circles and among the lay public. Ben-David did not share in it. Ostensibly, there were good reasons for the confident position with which he disagreed: at least 25 percent of Nobel Prize winners in Germany, when it was the leading scientific nation of the world, were Jews; when the United States emerged as the scientific superpower, Jews were winning at least 25 percent of the Nobel prizes going to Americans. It was sensible to predict that, at last in their own country, Jews would be doing even better. But Ben-David did not think so: intellectual talent, he argued, though a necessary condition for the advancement of knowledge, was not sufficient. To express and realize itself it needed a special environment. A small, tight-knit, undifferentiated community, in which each member performed a multitude of roles, personal and professional, vis-à-vis others—such as Israeli scientific community would be, if isolated from the wider scientific world—was not such an environment. In a tight-knit isolated community, personal and professional at once, independence of mind essential for scientific innovation would be in other respects counterproductive and originality would be likely suppressed (including self-suppressed). Special talent could assert itself only at the price of group harmony and would be punished at least as often as fostered. Scientific progress needed an impersonal, pluralistic, open environment in which ideas could compete for survival in a manner similar to the competition of genes in biology.[3]

The Israeli scientific establishment, Ben-David insisted, thus would not become a scientific superpower and the leader in scientific progress in its own right. It had to abandon the dream of becoming the scientific establishment of natural genius—for genius, precisely because natural, was the quality of individuals, not of nations—and reconcile with the fact that the great majority of scientists in Israel, by definition and like everywhere else, would be only "mediocre." The important aspiration, in fact, was to assure that this establishment be competent—mediocrity and competence, he argued, can very well go hand in hand—which was the best a scientific establishment (in distinction to individual scientists) could be. Ordinary, everyday science, called "normal" by Thomas Kuhn, science that goes on between breakthroughs and "scientific revolutions," is, of course, mediocre by definition. It is not "normal," but "great," "revolutionary" science, i.e., genius, that is the motor of scientific progress. But in a competent scientific establishment, normal science assures an informed audience for genius and makes possible the development of the ideas of genius. The competence of the scientific community, in other words, is a necessary condition for scientific progress.

Ben-David argued that the competence of the Israeli scientific establishment depended on its regarding itself as a part, a periphery, of nations whose resources allowed them to be true scientific superpowers, specifically, the United States. His suggestions, whether willingly or, more probably, under the pressure of structural factors he postulated, were implemented: The Israeli scientific establishment indeed developed into one of the most competent in the world, a periphery of the American network of research universities, but a periphery constantly and saliently contributing to the achievements at the center. Though Israel did not become a scientific superpower in its own right, Israeli-born and/or trained scientists are clearly over-represented among scientific super-achievers, but it is the pluralistic, ramified, and very large American system which allows them to achieve. Thus, his prediction proved accurate.

"Science in a Small Country" was, obviously, polemical, and many Israeli patriots disagreed with and were offended by it. Some still disagree: that Ben-David was wrong is proven for them, first of all, by the fact that Israel has emerged as a world leader, an indisputable center of innovation, in applied science or technology.[4] But, clearly, science in which Ben-David was interested was not, in this essay and throughout his career, applied science. It was "normal science" as an extremely important element of the environment in which scientific ideas developed (or not) and "first-rate," great science which consisted of colossal, fundamental, thought-changing breakthroughs, ideas that represented neither the application, however ingenious, of previously voiced ideas, nor the solution, however original, to the puzzles inspired by them, but a radical change of direction of inquiry—what Thomas Kuhn would call "scientific revolutions"—which we associate with the names of Newton, Darwin, and Einstein. One could not be trained to produce such science and even Nobel Prizes, the only means invented for its recognition, could not reflect it. It was this kind of scientific creativity—a most powerful and most disciplined imagination unerringly directed to the institutional goal of science, objective knowledge about the world of our experience, capable of breaking the mold of tradition and of initiating a new tradition, and all the while reined in by the principles of the scientific method of logically formulated conjectures (hypotheses) and refutations (the testing of these hypotheses against empirical evidence)—that Ben-David believed responsible for the progress of science. A competent scientific establishment encouraged such creativity (and an incompetent one discouraged it), but, in general, no social environment, while a central question for the sociology of science, could generate it. Its contributing causes came from the cultural universe outside the purview of sociology focused—especially as it was during Ben-David's lifetime—on non-symbolic social "structures" and their functions.

Modern science, both "normal" and great, is *empirical* science: not only does it have as its goal the understanding of experiential reality, but it rests its authority, ultimately, on experiential evidence. It is this dependence on the empirical which, in favorable social conditions, shapes the nature of continuous scientific activity as an institution and the progressive trajectory of scientific development. Such science, therefore, could only emerge when the experiential world was elevated to the dignity of the transcendental one (which it was soon to replace altogether) in the seventeenth century. However, some societies—China, for instance—have always focused on the empirical reality; why didn't science emerge there millennia ago, but had to wait for nationalism to appear in the West? This had to do with the other pillar of modern science, a peculiar invention of Western civilization, which had pre-existed the focus on the empirical by two thousand years—logic.

The centrality of logic for science is the subject of essays by Brendan Dooley and Dmitri Panchenko. It was the logical method of reasoning from first principles, argues Dooley, that had attracted some of the forerunners of modern science, such as Francis Bacon and Johannes Kepler, to astrology—the system of axioms about occult forces, which reflected a forgotten world of similarities and correspondences, symbols and clues to God's providence, and which now no one would view as a path to truth about nature. Yet, for those for whom nature was just a small sector of the vast transcendental realm, astrology must have been the very queen of sciences, as authoritative as theoretical physics is today. The method of reasoning from first principles, which, in the absence of the focus on the empirical, would make astrology as proper a way of scientific inquiry as any, and without which empiricism would be considered, as it was in early modern medicine, a method proper only for quacks, however, is not hardwired into the human brain, but is a historical phenomenon. Panchenko, the eminent Russian scholar of Ancient Greek science, refers to its origins in the sixth century BC Ionian city of Miletus as the emergence of "theoretical science." This he equates with "a tradition of argumentative knowledge, of improving and advancing knowledge through criticism and demonstration," and it is the establishment of this tradition that he attempts to explain. But for a tradition to exist, there, first, must be something specific to *tradit*—a tradition is a transmission of some symbolic property; it can only originate as a result of the emergence of this property. Only explaining the latter can provide us with a causal account of the tradition in question, i.e., with an answer to the question "why it emerged?" Focusing on the *conditions* of its establishment, in distinction, cannot lead to such an explanation. In the absence of these conditions, the tradition might not have been established, but they could not have brought it into being: conditions are not causes.

The radical difference in thinking brought about by "theoretical science" is recognized in referring to the period of its appearance (which is common among students of Ancient Greece) as the period of "transition from *mythos* to *logos*," while the difficulty of accounting for this transition is reflected in another common among classicists epithet: "the Greek miracle." It seems miraculous, and therefore unaccountable, because we believe firmly that logic is a natural human endowment, requiring no historical explanation, and what we have to account for is why it remained latent until the sixth century BC, a negative, which cannot be causally explained. But, in fact, logic is not a natural faculty, but a cultural creation, a product of someone's powerful imagination, which, like all such products, emerges in a particular historical context. It is logic that the tradition of argumentative knowledge, established in Miletus, perhaps thanks to the specific social conditions described by Panchenko, transmitted. Because we implicitly recognize that logic is absolutely fundamental to science, we consider the originator of this tradition, Thales, the father of science, but so far our belief in the universality of logic has prevented us from focusing on him as the progenitor of logic itself. Thales was the first "first-rate" scientist of the kind Ben-David had in mind in his admonition to the young Israeli scientific establishment: he broke the mold of the mythical tradition and founded the logical one. Only a very special image of reality can accommodate the principle of no contradiction and suggest logic as a way of inquiry into it. (It is quite suggestive, I would point out, that "the Greek miracle" happens in Asia Minor precisely at the time when Jews in exile in the adjacent Babylon are working on the first redaction of the Hebrew Bible—the very first systematic presentation of monotheism.)[5]

Ben-David was not particularly interested in the question why and how modern science emerged, why in a certain place and not elsewhere, why when it did and not at some other time. The most important question for him as a sociologist of science was the question of the structural conditions conducive to the development of science after its twin pillars—logic, and the focus and dependence on the empirical world (the twin foundations of both the scientific method and the normative structure of science as a social institution)—were firmly in place. For a long time, Ben-David's main claim in this respect, put forth in the famous article he coauthored with his student Awraham Zloczower,[6] was interpreted as the argument that science was best (i.e., advanced most speedily and led to the deepest objective knowledge) when competition between composing units of the scientific establishment (i.e., universities) or between individual scientists and groups of scientists was encouraged. As both Claudius Gellert and Mario Coccia point out, if interpreted this way, the claim does not hold true, or, at least, no longer holds true today. Gellert examines the original German case, on which the Ben-David–Zloczower study was based, and uses the American and English cases for comparison, while Coccia chooses Italy as the focus of his detailed analysis, but adduces sufficient data to make it evident that the Italian situation is quite representative of other Western European cases, most importantly showing a great similarity to France. Great European nations—especially those which at some point in the past few centuries were undisputed centers of science—find it difficult to reconcile with the position of a scientific periphery. Based on the common, literal interpretation of the claim that competitiveness of a national scientific establishment (i.e., the standing of a nation's science vis-à-vis scientific achievements of other nations) depends on the level of competition within that establishment, governments of these nations in various ways attempt to foster the competitive spirit among, and, in effect, force competition on, their research institutions and scientists. Time and energy of people involved in science are consumed in competing for funds (public or private) simply to ensure the continued existence of the institutions and employment of individuals. Fundraising ability is made the foundation of scientific careers, and scientific talent is wasted. Both Gellert and Coccia show that such enforcement of competition—which amounts to the organization from above, regulation, management, or dirigisme of scientific research—does not lead even to an increased productivity (meaning quantity of publications) among the competitors, and may be directly counterproductive of the progress of science, that is, the quality of knowledge produced.

Was Ben-David wrong in suggesting that unrestrained competition within a system was the source of scientific progress? No, it is the interpretation of the competition he spoke of as social competition (for money or status) that is wrong. Clearly, what he had in mind is competition far more akin to the natural, Darwinian one, between ideas for survival. For such competition to occur, society, like nature, has to be free, unregulated, uncentralized, unmanaged. The essential condition for scientific progress never was competition as such, it was the autonomy of science. It was because the nineteenth century German university system, itself a product of the ramshackle, feudal, fundamentally decentralized political character of German society, united only by language, allowed the autonomy of science—specifically, its autonomy from economic and political pressures—and because the twentieth century American research university with its graduate departments safeguarded this autonomy, that, Ben-David believed, science flourished in these two societies, otherwise so different in their structure and political culture.

Two papers in this volume show how absolutely central for Ben-David was the autonomy of science: those of Chikako Takeishi and Yaron Ezrahi. It was not just a condition *sine qua non* for the progress of science, but, when consciously promoted (as he thought it was in America) a supreme moral value, an expression of the just, liberal democratic society and political order. It was, thus, both an idea and an ideal, which made it difficult for some to distinguish where sociological discourse ended and a moral discourse began. Science, for Ben-David, clearly had a social role of paramount significance, but it was not the role of national aggrandizement or that of an economic prop. It was not a utilitarian role at all. The greatest service science rendered to society was that of an example; it served mankind best by simply being. The paper by Chikako Takeishi makes this evident in a persuasively descriptive, non-argumentative way. Takeishi discusses a very important but little known aspect of Ben-David's legacy: his criticism and recommendations for reform of the Japanese scientific/high education establishment. Because of the nature of his involvement in Japan, his ideals, especially the ideal of the university, are very well reflected in this criticism and recommendations. During his visit to Japan in 1970 as one of the five members of the OECD committee sent to the country to examine its national policies in education, Ben-David found Japanese universities wanting. They were too focused on "social selection," constituting "a single hierarchy of prestige," and this focus was an obstacle to the achievement of educational (training) diversity, essential for the promotion of scientific excellence. They were also not "autonomous" enough: he was surprised, for instance, that university reforms were initiated by the government rather than by universities themselves. The two desiderata expressed the ideal of academic freedom, which the American university apparently embodied—that is, freedom to pursue one's research in pure science wherever it leads one and freedom to teach the conclusions to which such unconstrained research has led, or the autonomy of science by other name. The university was, for Ben-David, the home of pure science, whose objective was to increase humanity's understanding of empirical reality. Its role was not to foster applied science, which, by definition, served the needs of specific customers (whether government or private sector) and which, therefore, exchanged academic freedom for financial support. The university was to pursue its end unswervingly and to resist the pressures (and the temptations) of the government and other consumers of science. Support of pure science within the university, Ben-David wrote in his Japanese report, could not be based on "possible benefits to the economy or to the solution of a practical problem," and it could not be evaluated by its consumers, but only by the peers of the pure/academic scientist in question.

Increasing humanity's understanding of empirical reality being dependent on the development of the ability to understand, i.e., development of the mind, in general, the university, as the home of pure science, was entrusted with a major educational responsibility, which was fundamentally different from the responsibility to train future researchers and professionals. In the United States, the emphasis on liberal arts education was a survival from its college era, the most prestigious research universities—Harvard, Yale, Princeton, Columbia—all being reformed liberal arts colleges. But, clearly, the need for general education, which would develop one's mind rather than simply train one for a specific line of work, was great on the American campus with which Ben-David was familiar, and he found the same need—unanswered—in Japan as well. With increasing democratization and college education becoming the expectation of broad social strata, the numbers of students going to college to "find themselves" increased, thus increasing the university's general educational responsibility. "Devoting serious attention to the problem of moral education at the universities," Ben-David argued in his recommendations to Japanese academic establishment, "may not only help students and the universities, but also society." Such general education, however, had to be "a pluralistic enterprise," and not "uniform political indoctrination," as it would be, "if the programs are contrived uniformly according to some central directive." It was natural to think, in the midst of the Cold War, that uniform political indoctrination was always a result of some central directive.

Today Ben-David's criticisms of the Japanese scientific and tertiary education system apply as much, if not more, to the American research university, which he saw as the model to be imitated. Clearly today (and for the last quarter-century, at the very least) American universities are focused on social selection and constitute a single hierarchy of prestige, which precludes diversity in scientific training and discourages scientific innovation. American universities also completely gave up their "autonomy": the distinctions between pure and applied science (science and technology, in other words) were blurred and, with promotion and professional status, in general, dependent on outside grants awarded, research is shaped—in its questions and answers—by the requirements of funding agencies, governmental and private, interested only in the solution of practical problems. This thorny issue, already addressed in the papers by Gellert and Coccia, is the focus of Yaron Ezrahi's impassioned essay in the volume. It is the expansion of the market forces, in particular, into science, which, in Ezrahi's view, is mainly to blame for the loss of the autonomy of science: science, he claims, has become a mercantile, capitalistic enterprise, as unlike Ben-David's spiritual ideal as can be. But what allowed the market forces in, the ultimate, rather than only proximate, cause of this terrible loss, is, of course, far more important, and it is none other than the loss of self-confidence, faith in itself, on the part of the liberal-democratic society as such. Autonomous, free science was the centerpiece of the enlightenment vision at the core of the moral structure (and political culture) in Western democracies. Bringing this entire value order increasingly into question, in fact replacing it with what Ezrahi calls "moral perspectivism," inevitably eroded the support for science among the lay public and rendered even scientists, confused as to the nature and dignity of their activity, incapable of protecting its autonomy.

All this applies to science in general, across the disciplinary boundaries and irrespective of the subject matter. But social sciences face even greater problems, for even in that "golden age," when science was indeed allowed to enjoy its full autonomy, they remained stagnant and failed to advance our understanding of human reality. The main reason for this seems to be the absence of the possibility of imagining human reality as an autonomous subject regarding which objective knowledge could at all be generated, the essential cultural precondition for scientific creativity (or "first-rate science") in any area, which Ben-David left out of his purview. Ben-David's fundamental position in the sociology of science was that structural factors had no causal influence on the nature of scientific ideas and could only explain why certain ideas succeeded in certain circumstances. In his writing, as a sociologist, he concentrated on the structural factors—sociology could not contribute anything else, in his view, to the explanation of scientific progress. But this focus created an impression among those familiar with his thought only through his books and major sociological articles that he endowed conditions for "normal science" with power to generate revolutionary scientific ideas. Sometimes, curiously, in unguarded moments, he seemed to have shared in this impression himself. But this was, obviously, a contradiction. For example, it is easy to interpret Ben-David's discussion of the American research university system as a reflection of a belief in its capacity to actually foster and generate scientific progress. It could not do so; of course, no bureaucratic system can do more than foster competence in doing things in an already established way: this explains why all the real innovators in science were academic outsiders (Einstein) or altogether unfamiliar with the bureaucratic yoke (Newton, Darwin). Ben-David also believed in sociology and thus accepted the division into disciplines in the social sciences on the whole uncritically, probably thinking that these disciplines followed the developmental trajectory of subdisciplines in physics and biology.[7] Clearly, for him, the American research university which allowed the "normal" physical and biological sciences to ramify prodigiously would allow similar growth in the knowledge of human society and, in general, constituted the optimal conditions for scientific progress.

It is becoming increasingly obvious that the compartmentalization (or departmentalization) of the study of human society[8] has been one of the main obstacles to the progress of knowledge in this area. As Nathalie Richard demonstrates in this volume's essay, in one very important instance departmentalization indeed precluded the development of a tradition which could have led to a true *science* (empirical, logical, and progressively generating objective knowledge) of humanity. The idea of such a science—*science de l'humanité*, or *science de l'homme*—was simultaneously espoused by three French scholars,

Hippolyte Taine, Ernest Renan, and Alfred Maury, in the nineteenth century. Unfortunately, conditions for a tradition that would transmit it did not exist. The reform of the French universities on the recently established American model with its specialized departments, which obstructed the fact that, like matter in all the physical sciences and life in all the biological ones, the central subject in all the social sciences and humanities was, indeed, humanity, led French scholars in other, clearly less productive, directions. Taine's, Renan's, and Maury's "science of man" recognized mental activity as humanity's distinguishing characteristic which should be the focus of its study; it was, in this sense, very innovative and contained a tantalizing promise. It is not certain, of course, that it would lead to the advancement of knowledge: their resolute psychologism was biologistic, and their imagination was limited by the age-old ontological view of objective reality as consisting of two substances, material, which they considered fundamental, and spiritual, which they believed epiphenomenal. But, unquestionably, there was a chance of such advancement.[9]

The situation in regard to creativity in all the areas of science, when the activity as such no longer enjoys the necessary for it autonomy, is not helped by the tendency of "normal science" to regard the currently dominant paradigm as the last and the only correct word and to stifle creativity through the very peer review that Ben-David considered its main safeguard and by the tendency of bureaucratic organizations to breed mediocrity in its emphasis on competence as the only standard of excellence. Yet, this is not the most dismal, the gravest and the most dangerous in its consequences, effect of the devaluation of science and ideological attack on it and the liberal values it represents. That pride of place belongs to the failure of the university in the general, "moral," education of college students with which it has been entrusted. Tyranny of the majority (of the faculty) on campus has proved as potent a factor in ensuring uniform political indoctrination as any central directive. How such indoctrination substitutes for general education all across American universities is the subject of the meticulously documented and, for this reason, consistently shocking paper by Peter Wood. Though, obviously, there are (and, one hopes, numerous) exceptions to his depressing account, according to Wood's data, the tables have been turned: the American campus in 2010 feels nothing so much as the chilling world of 1984, which it would be difficult to imagine in today's liberalized China or Russia. It is a *totalitarian* environment precisely in the sense the inventor of the term, Benito Mussolini, gave it—it recognizes no privacy, no right to act, feel, or think on one's own, leaving the individual no corner to hide from the judgment, and to escape the punishment, of the authorities presenting themselves as spokesmen for the will of the human *and natural* world. Indeed, the arrogance of both Soviet and Chinese communist leadership, whose existential attitudes and psychological brutality these authorities—faculty and administrators—of the American research universities so obviously adopt, fades in comparison to the latter's sense of self-importance: after all, the Communists claimed to represent only the will of humanity. But they had to deal mostly with adults, and there is a limit to the credulity of even the most credulous, mentally laziest adults, and even the most cowardly of them may be expected to revolt when this limit is disregarded, while the powers at the American research university have as their flock eighteen- and nineteen-year-old students, veritable children. These children are a captive audience, their parents paying through the nose for the certificates the universities are expected to bestow on them for compliance, and absolutely defenseless against the indoctrination practiced on them by those on whom getting these exorbitantly expensive certificates depends, given that their minds are yet unformed. The university, therefore, actually becomes a "total institution" in Goffman's sense of the term—a prison for the young minds, a correctional facility—not for juvenile delinquents, but for white girls and especially boys, judged to be of a congenitally imperfect, in fact inherently wicked, racial stock. For, ironically, though the American university today, and for the last half a century, is identified with the ideological left, their actual ideology, race- rather than economics-based, is clearly more reminiscent of the paradigm on the right.[10]

This is not surprising: this ideology is today's form the Romantic mentality, whose first and most pungent fruit was German nationalism, takes in the United States. Speeches and publications of our homebred totalitarians in environmental clothing, which Wood tirelessly quotes, reverberate with evocations of creative emotion vs. stilted reason; of the vital superiority of the group vs. the individual—the submergence of one's individuality in the group, of the right kind, obviously, being both natural and morally imperative, and, interestingly, the only way of achieving real individuality; of the natural, and therefore morally superior, nature of women and "minority cultures" vs. the artificial, and therefore wicked, nature of men and their white or Western "culture"; and of the comforts of ascription vs. the dangers—and moral abomination—of individual achievement. Also notable is a hostile attitude to science, that very embodiment of everything that is unnatural and evil in white men and Western "culture." All this is found, often word for word, in the opus of proto-Romantic German authors of the *Sturm-und-Drang* period, that of early German Romantics and neo-Romantics of the early twentieth century.

The immediate preoccupations of the Cold War obscured the deeper dynamics underneath it (among other things leading to the identification of anti-Western politics with the ideological left). That is why the whiff of the *animal farm* Ben-David discerned on American campuses already in the 1970s appeared to him to come from that direction, and to originate outside of the American society in the centrally controlled states. But the real conflict (as opposed to the form it took between 1950 and 1990) was of a much longer duration, beginning in the end of the eighteenth century, and was essentially internal to every society that has experienced it. This is the conflict between the two core values of modern democratic society: liberty and equality, the conflict between those who accept the anomic insecurity of the free society and take freedom as an exciting challenge and those, fearful of failure and envious of the success of the risk-takers, who want their equality to be assured and who, therefore, reject free society because of the anomic insecurity it implies. It is always the sense of one's own inferiority that lies behind fear of failure and envy of those who succeed—a psychologically crippling, enduring emotional condition that cries out for therapy. Articulation, and especially imposition on others, of the belief that one is not inferior but is only made to feel inferior by the evil dominant "culture," first accomplished by the Romantic authors in Germany, has been the preferred variety of such self-therapy. It is now entrenched in the American research university—which makes it an unlikely structure to promote values Ben-David considered dear: scientific creativity and sense of clear identity among undergraduates coming to college to "find themselves." Without any individual Pig emerging as the Leader, our campuses have become branches of Orwellian *animal farm*.

Michael Ben-Chaim adds to the indictment of today's university as the leading educational institution an extremely important note: he points to the fact that educational philosophies developed in the university and reflecting its self-understanding (and the self-understanding of science) directly, profoundly, and inevitably affect the education of children from kindergarten on. A historian and philosopher of science, but also a practicing educator on the secondary level, he, like Ben-David (whose very last doctoral student he was), believes that the goal of education is the development of the student's mind, helping one to realize one's humanity; like Ben-David in his recommendations to the Japanese scientific/educational establishment, he considers intellectual diversity and autonomy to be the foundation of the development of the mind (the ultimate source of creativity in science, among other things). But he also believes that the university has long substituted for such education the imparting of (ostensibly true) information in ways that undermine the cultivation of the students' critical thinking. This, he claims, is a result of the *scientistic* ideology, i.e., the widely accepted view in modern society that current scientific knowledge is the only knowledge that is real and worth knowing, in general. This general ideology has been propagated by the modern—scientific—university since the days of Humboldt, imposing conformity to established standards and, thereby, necessarily constricting diversity and autonomy. This is, of course, correct: intellectual conformism is antithetical to creativity, by definition; that is why great innovators come from among outsiders rather than from among competent specialists bred within bureaucratic establishments. However, if even the ideology of science insisting on the dedication to diversity and autonomy has such deleterious effects on diversity and autonomy, how much worse must be the effects of university-propagated ideologies which do not care for diversity of ideas and deprive science of its autonomy!

The reasons Ben-David considered the American research university the model for scientific and higher education institutions in other countries no longer exist, though it still serves as such a model—such is the power of social inertia. But Ben-David's ideas about science and universities help us to understand what is happening. Their continuous relevance, now, twenty-six years after his death, is a testimony of extraordinary achievement. The annotated photographs, around which Ben-David's widow, Miriam, organized her personal talk at the conference, and which serve as the epilogue to this volume, chart Ben-David's life from childhood in Hungary of the 1930s through the Holocaust's devastation of his family to his recreation of himself as a social scientist, and place his ideas and ideals in the context of his personal experience. Specifically, they help one to understand his fundamental belief that the university represents the bastion of civility in society, his abiding concern that, if the rational discourse, unencumbered by prejudices and uncontrolled sentiments, cannot take place within its walls, there is no hope it could find a place anywhere else. This volume demonstrates our commitment to this view. United in our appreciation of Ben-David's work, we obviously disagree on many other issues. Our ability to see this project from inception to completion, however, proves that in the community of scholars it is quite possible to disagree with civility, rather than acrimony, to disagree and yet cooperate productively.

Notes

1. See, for instance, Benjamin Ginsberg, *The Fall of the Faculty: The Rise of the All-Administrative University and Why It Matters* (New York: Oxford University Press, 2011) and "Faculty Fallout," the-scientist.com.
2. See "Science and National Greatness in 17th Century England," in Greenfeld, *Nationalism and the Mind: Essays on Modern Culture* (Oxford: Oneworld, 2006), this was the last paper I discussed with Joseph Ben-David, and it was first published in the special issue of *Minerva*, dedicated to his memory, just months after his death; Thomas Spratt, *The History of the Royal Society*, London, 1666.
3. In both contexts, obviously, the word "competition" is a metaphor, because in neither we are talking about actual, much less intentional, competition between individuals. Genes in biology survive thanks to an adaptive niche in which they are not eliminated; similarly, ideas as if offer themselves for sale in an anonymous marketplace and find niches where they can develop.
4. See George Gilder, *The Israel Test* (Minneapolis, MN: Richard Vigilante Books, 2009); Dan Senor and Saul Singer, *Start-Up Nation: The Story of Israel's Economic Miracle* (New York: Twelve, 2009).
5. The principle of no contradiction, on which Aristotelian logic is based, is *logically* impossible in the framework of polytheism. The first systematic presentation (and discussion) of monotheism is, therefore, the only way to account for the "miraculous" transition from *mythos* to *logos*. (See my *Mind, Modernity, Madness: The Impact of Culture on Human Experience*, forthcoming.)
6. Joseph Ben-David and A. Zloczower, "The Idea of the University and the Academic Market Place," *European Journal of Sociology* 2 (December 1961): 303–14 and "Universities and Academic Systems in Modern Societies," *European Journal of Sociology* 3 (June 1962): 45–84.
7. They did not. See my "How Economics Became a Science: A Surprising Career of a Model Discipline," in *Disciplinarity at the Fin de Siècle*, ed. Amanda Anderson and J. Valente (Baltimore, MD: Johns Hopkins University Press, 2001), 87–125.
8. Into the study of human society through social relations in various spheres of life in the West at present (sociology), the study of human society through social relations mostly in the West in the past (history), the study of human society through social relations pertaining to the generation and allocation of material resources mostly in the West mostly in the immediate present (economics), the study of human society through social relations pertaining to power competition between different national groups and between

subgroups within nations mostly in the West mostly in the immediate present (political science), the study of human society through social relations in non-Western societies at present (anthropology)—one needs only define the separate subject matters of these numerous "disciplines" for the absurdity of such division to become starkly apparent.

9. For an extended discussion of departments as a structural obstacle to the advance of the social sciences and of the dualist ontology as an intellectual obstacle, see my essay, "American Universities and the Stagnation of Knowledge" in this volume.

10. Indeed, on some campuses, open anti-Semitism, even violent, is now considered politically correct. The next logical step could be the burning of books. Jews who participate in this totalitarian Bacchanalia should beware: once one starts the ball rolling, it might be very difficult to stop it. (See communication of April 21, 2011, from Carry Nelson, AAUP President, regarding "Anti-Semitism on Campus," AAUP newsletters.)

1

Science in a Small Country

Joseph Ben-David[1]

As a widely accepted opinion has it, Israel, although a small country, is a mighty power in the scientific realm or, at least, it shall become one in the foreseeable future. This opinion rests on two assumptions which remain unexamined and unconfirmed. According to the first, the Jewish people are the people of the book, dedicated to a long tradition of study and inquiry, who in each and every generation have brought forth great scholars. Thus, it is to be expected that such a people will rise to even greater heights in their own country. Intellectual talent, indeed, is an essential condition for scientific creation, yet it is not a sufficient condition. Inasmuch as we have been throughout history the "People of the Book," we have been the people of one single book, or of a limited number of books, all concerned with that same book. We have studied nothing but the Torah and we may have achieved much in this domain, but it is impossible to evaluate our greatness for we have been sole competitors in this field. Those great rabbis born to the Jewish people throughout its long history may have been geniuses, but they were not scientists, nor even scholars who are part of the general history of thought. One could claim that in the period beginning with Jesus or, possibly, with Philo of Alexandria, and ending with Spinoza, the Jewish people did not create any first-rate work to have transcended Judaism and to have had any impact beyond the Jewish community. Moreover, both Jesus and Spinoza were, to put it mildly, marginal Jews, and insofar as they have achieved major achievements these were achieved despite their Jewishness and in stark opposition to the Jewish communities of their respective times. When we look at modern science, whatever may be the meaning we ascribe to this term, we are bound to conclude that the history of science up to the second half of the nineteenth century would not have taken a different course had there not been Jews in the world.

The second assumption underscores Jewish contributions to science since the nineteenth century until the present. In this period, it is claimed, Jews have done much in the sciences, the humanities, as well as in art and literature. For instance, the percentage of Jews among Nobel Prize winners is the highest in the world: about a fourth of Nobel Prize winners in Germany were Jews, and as many, if not more, among American Nobel Prize winners are Jewish. However, we need to tread with caution when we come to infer from these happy facts anything regarding our future prospects in Israel. All the Nobel Prizes obtained—and not only the Nobel Prizes which, after all, are but a token of external appreciation—and all the outstanding scientific works carried out by Jews in all scientific disciplines were obtained and achieved abroad. Here, in Israel, Jews haven't made any scientific breakthrough, and those first-rate scientists who may be dwelling among us are all immigrants from abroad where they were first recognized as leading figures in their fields. Thus, we have no evidence whatsoever to support our presumptions and our convictions that we stand more than a fair chance of becoming a scientific society soon.

One could counter my affirmations by saying that whatever data I am adducing, my conclusions could have been more optimistic, and that we are entitled to project from the achievements of Jews abroad onto our possibilities in Israel. If we haven't accomplished more as yet, it is only because we are a young, struggling state, and so on and so forth, and time will tell. To this argument I will retort that the majority of those outstanding Jewish scientists abroad were assimilated Jews, that is, people who were, culturally speaking, Americans, Germans, Frenchmen, and Englishmen, albeit of Jewish descent. I need to underline, time and again, the scientific wilderness where Jews wandered for long years down to the middle of the nineteenth century. Indeed, it is very unlikely that we will encounter such desert expanses again for the wilderness was the outcome of a conservative theocratic regime that sharply condemned scientific pursuits whereas today we are no longer subject to such a regime. Nevertheless, although Israel is not a traditional theocratic state, long-standing religious influences still shape it, and a sizable portion of its population believes that the Vilna Gaon was by far a more important figure than Einstein. I have no intention of undermining their belief, and they may very well be right. All I can say is that as far as this sector is concerned, one can expect it to spawn new towering figures in the mold of Rabbi Elijah of Vilnius rather than new Einsteins.

Even when we look at the general population of Israel, we shall see that a number of problems arise by dint of our being Jews in Israel: one of the characteristics of any first-rate creative work, be it in the sciences or in the arts, is individualism which at times will manifest itself in cruel and extreme forms.

A person bent on creative work cannot be bothered with other matters, and should he get engrossed in matters extraneous to his work, it is doubtful he will achieve anything great. This doesn't by any means imply that a creative person is not an idealist aspiring to benefit humanity at large by his scientific discoveries. In a few exceptional cases, it is possible to establish a causal link between a discovery and a desire to contribute to the welfare of humanity. Yet what is usually required of the creative person is mastery of his work and nothing else. One of the reasons that might account for Jewish creativity abroad is the relative marginality of Jews in relation to society. As marginals, Jews could afford being more selfish and more individualistic than others. The society they dwelt in was not their society, and they could allow themselves to retreat from it and enclose themselves in their chosen creative fields. In Israel, the state of affairs is markedly different: here we live under constant ideological collectivist pressure that by its very nature goes against the grain of scientific endeavor. As an example, I can cite an interview I have conducted a few years ago, when I was studying youth movement members, with the son of one of our professors. This young man confessed that his father convinced a new immigrant who became a kibbutz member to leave his kibbutz, pursue his studies at a university, and devote himself to scientific work, and described his father's actions as immoral and shameful. By the way, the man who followed the advice of that father who brought so much shame upon his son turned into one of the top Israeli scientists in this discipline, an invaluable addition to the local scientific scene. Only seldom had nonobservant Jews abroad to cope with such a situation, with a conflict that stands in the way of creativity and clips its wings. And the example I brought from science can be easily complemented with examples from the history of literature. As a matter of fact, the body of writing usually referred to by the name "the new Hebrew literature" which was written between 1945 and 1955 is hardly worthy of its name, among other reasons because the authors who are responsible for it were not capable of writing about flesh-and-blood human beings but only of Palmach fighters. The collective picture, so to speak, overshadowed and suppressed true creation, because collective problems lay beyond the scope of creativity. It appears, then, that Judaism, which in the past was a factor of selection that shaped Jews as a particularly gifted people, possibly as more gifted than any other, cannot at present in Israel guarantee any future scientific or artistic greatness.

Let's suppose now that if we were to weigh our hereditary gifts against the problems I have mentioned, we would not fare worse than other small nations. It is even possible that we will fare much better. After all, among us the rate of people with higher education is particularly significant, and this fact probably attests to some talent. And yet, the question remains: to what extent can a small country achieve greatness in science?

Here, too, or so it seems to me, only a few are aware of realities and are willing to look at them squarely. Small countries need to cope with specific problems which stand in their way, and that applies to all small countries and not only to ours although these problems are especially severe in Israel due to its size. One of these problems is scientific excellence. Science is dynamic, constantly evolving. One could say that the majority of renowned scientists, that is, people who have already marked their scientific fields, belong to the past. They will not add much to their previous achievements. The greatness of science, or of the scientific system of any given society, is conditioned by dynamism, and by the existence of a circuit of people who will not content themselves with their previous achievements but rather consider them as stepping stones to the next stage and to the not-yet-achieved. In order to create dynamic standards that impose on the scientific community ever-growing tasks, it is necessary to have a specific type of social structure, which is founded on the constant selection of the first-rate as opposed to the merely efficient. Sometimes big countries encourage such a selection. Only in big countries can there exist a competition among institutions, which enables some of the institutions to become truly first-rate. By "first-rate institutions" I refer not to a place that accidentally hosts an exceptional individual while the rest of its staff are merely dedicated people who know their jobs, but to an institution where a considerable portion of the scientific staff consists of truly creative people who congregate there in order to enjoy the particular atmosphere of a scientific forefront. Only a constant selection of the exceptional can lead to the creation of such institutions.

In a small country all this is well nigh impossible. Statistically speaking, there is little chance that in a country as small as Israel, and even in a country ten times bigger, there will emerge great scientists. It needs to be emphasized that I am not speaking here of good scientists of which there are many. I am speaking of people capable of producing original work, and their number, in whichever field, is of necessity small. Should I predict that at any given moment we are likely to have among us first-rate scientists in five distinct disciplines, I will be overstating the case. Yet even if that were ever to happen, it would imply that in all other disciplines and specializations we would have only mediocre scientists. There is no shame in it, for "mediocre" people can be worthy scientists, but they are not original thinkers. Nevertheless, when mediocre scientists are the only scientists in their country, it is they who dictate the professional standards of their respective disciplines. Local standards, then, can be internationally high in a discipline or two, but in all other disciplines the standards will be low from the international point of view, although locally, mediocre standards will constitute the highest standards.

When standards of mediocrity emerge in a small country, and when these standards are translated into influential positions, it is very unlikely that mediocrity will ever be transcended. Mediocre practitioners usually enjoy monopoly, and all the more so when there is only a single position or a single department in any given field in an academic institution. Even if there were a number of departments and positions, practitioners would end by forming a cartel and imposing the standards they see fit. Far be it from me to accuse anyone of deliberately lowering requirements or suppressing creativity. However, we need to admit that the potential of creative thinkers often goes unrecognized. At times, a ten-year-old shows great musical or mathematical promise (and, indeed, music and mathematics are the only fields where genius manifests itself at a young age and is often recognized early), but these are exceptions. As a matter of rule, there are doubts. Young original scientists are often attracted to what may appear at first glance as marginal, dubious, and disreputable, and that is so for obvious reasons. The "respectable" is that which already exists and, therefore, it is less appealing as there is little that one could create there. In order to create the new, one needs to take risks. Yet those who lack vision often perceive the risk-taking of the young as dubious pursuits. Unable to understand originality, they will attempt to draw it back into the bosom of mediocrity.

These statements concerning the rule of mediocrity are not abstract theory. It is a realistic description or, at least, a description of past reality. Indeed, in the 1920s and especially in the 1930s, Israel had "inherited" a great many world famous scientists who, for the most part, emigrated from Europe. By 1950, in science, Israel lagged far behind other countries. In certain disciplines, scientists from abroad were still active although here they added little to nothing to their past work. In other disciplines, the state of affairs was sub par, and standards far lower than mediocre reigned supreme. At that time, Israel was closed to all possible scientific contacts, and the mechanism described above, that pull toward mediocrity, was left unchecked. The country stood on the brink of scientific balkanization, and was saved only thanks to some drastic changes.

Another difficulty that stands in the way of scientific development in a small country—and in this respect we are no different than any other small country—arises from the following paradox: despite the smallness of the geographic size and the tightness of social networks, there is no scientific collaboration. Teamwork comparable to the teamwork one encounters in various fields in the United States cannot take place in a small country. Even mutually stimulating intellectual dialog is a rare occurrence. Often at international conferences, I and many others have encountered foreign colleagues who tell us a story that runs along these lines: "I have met an Israeli scientist who told me he's working on a problem and would like to have my lights on the question. When asked why he didn't consult such and such who is also Israeli and a well established authority on the matter, he simply replied, 'I had no idea.'" How can we account for that? By pointing to a fear of communication. Inasmuch as communication is established, it is rife with tension, and people do not seek it. Those who desire discussion need to go abroad, and it is there that they find new inspiration. "Scientific communities" simply do not form here, even in those fields where one would expect them. Scientific friendship is suspect. When someone happens to be my friend, I accept him as he is, whether clever or foolish, a success or a failure, and nothing of all the above can diminish or strengthen our friendship. That is not so in scientific friendship, where the driving force is not empathy but intellectual affinity. Scientific friendship is peculiar and particular. In a big country, a practitioner in any given scientific field participates in a group of hundreds or thousands of members with whom he is familiar by way of scientific literature or conferences, and whom he appreciates objectively on the grounds of their scientific work. Some of them he may even know personally. He lives in a "community" of people he knows only as scientists and judges objectively. They have neither harmed nor helped him. He can measure their true merit, and that applies also to his personal friends who are scientists, as it is clear to all that scientific appreciation differs in nature from individual appreciation. In a small country, things are different. All those involved in a certain discipline are personal acquaintances; each is either a teacher, a co-disciple, or a student of the other. Each has had the opportunity to help or harm the other. Thus, each practitioner forms a nexus in a thicket of personal relations of which the scientific aspect is but one and that aspect cannot be separated or demarcated from the others. As scientific relations need to be exclusive, human relationships among scientists tend to become complicated. A criticism, a refusal to collaborate, will soon be interpreted as a hostile gesture, and then explained as hatred and persecution. Small groups of colleagues give rise to cliques, alliances, tensions, and suspicions that cannot be distinguished from scientific activities. Anyone wishing to escape this turmoil will find himself working alone and isolated.

This is how things stand in the vast majority of small countries. Is there a way to escape the pattern, the isolation, and to overcome the limitations that pertain to our being a small country? We can point to a few historical cases where small countries prevailed over their limitations, at least for some time. Switzerland, for instance, was a great scientific country in the second half of the nineteenth century and in the beginning of the twentieth, although it has lost ground later on. The Scandinavian countries, too, withstood the test for a long time, although it is doubtful whether one can refer to them as small countries, as the contacts among them are close, each of them speaks each other's tongue, and they all coalesce into one reciprocal system. Another encouraging example is Holland, which has proved it was capable of considerable scientific accomplishment. If we ask ourselves how those accomplishments were achieved, we shall see that our considerations have direct bearing on the scientific policy we may wish to implement in our country, and that might prove adequate to its needs and capabilities. Switzerland, I believe, provides the clearest example, and that is so because we can trace with precision the moment of its rise and of its decline as a scientific power. It appears that in Switzerland at the time of its rise to prominence, scientific exploits were intimately linked to the fact that the Swiss German-speaking universities, and in particular the University of Zürich, formed an integral part of the German scientific system. These universities never sought to compete with the universities of Berlin or Vienna, Leipzig or Munich. As universities in a small country, they were aware of their possibilities and of their limitations. Whether deliberately or by accident, Zürich turned into "the antechamber to the first-class German academic system," as indeed it was often called. Many young scientists, of whom Einstein is but one example, accepted invitations to teach at the University of Zürich at an early stage of their careers, before they had made a name for themselves and could aspire to a position in one of Germany's hallowed scientific institutions. Having given up the attempt of becoming an independent center that competes with Germany and struggles to attract and hold on to famous scientists, Zürich learned how to turn necessity into a virtue. The scientists who came to Zürich might have stayed only five or ten years, then moved on to other universities, but that often meant that they spent their most fruitful teaching and research years there. From one point of view, Zürich was in a desperate and wretched situation, constantly losing its best scholars. On the other hand, precisely for this reason, it kept on evolving and advancing. If it were to accept scientists only on the condition that they would remain there permanently, it would have made no progress whatsoever. However, having agreed to becoming a part in a much bigger system and to sustaining losses, and, at the same time, having decided to experiment and to try to win as much as it could, Zürich launched a process of continuous development.

It seems to me that the line of thinking delineated above is the very opposite of our approach to our own scientific system. We think of Israeli science as of a thing apart, as a closed entity. Surely we are interested in contacts with the scientific work done abroad, but these contacts are subject to painstaking calculations: how many people do we "get" from abroad, and how many do we "lose" to foreign institutions? Each "loss" is cause for national mourning. Always afraid of such "losses," we turn to Jewish scientists abroad and stipulate: you ought to come to Israel, forge ahead with us, but only on the condition that you stay with us for good. If we insist on this condition, our chances of development will dwindle. We need to see reality for what it is. We need to accept the fact that we are but a part in a global scientific system, a scientific province, especially of the United States. That is so in most disciplines, and many Western countries, much bigger than us, are in a similar position. If we would only accept this fact, we no longer would have to tremble for every gifted Israeli student who spends more than five or seven years abroad. At the same time, this will also enable us to create here adequate conditions for foreign scientists who might wish to spend a limited amount of time in Israel and then move on to other places and greater possibilities. Such an approach entails risks, but it also yields great dividends.

It must be added that inasmuch as we have evolved, we have evolved thanks only to adopting precisely such an approach. We have sent students and scientists abroad; we have "lost" our people to foreign universities. We have turned Israel into a country scientists visit regularly, for longer or shorter periods (although all too often the visits are too brief and too ceremonial). The progress we have made here since 1949–50 was not due to some major scientific discovery or to successful competition with bigger countries, but only to the fact that we have entered the "bloodstream" of global science. We haven't established an Israeli science, but rather we have become a modest member of the international scientific community. And the achievements we have accomplished thanks to these measures by far surpass our achievements in the period of autarchy.

This has many implications not only in matters of hiring policy, but also in matters of research, that is, in matters of research problems and research methods. Officially, we still suffer from a *folie de grandeurs*. We want Israel to send rockets to the moon or, at least, to some great heights. Our newspapers still write all sorts of nonsense about Israel's entry into the rocket club. I have no intention of pronouncing myself on the political and defense aspects of this approach, but from the point of view of scientific creativity it is entirely wrong. A certain kind of research requires funds, manpower, and long-term planning. Needless to say, a small country is in no position to meet those requirements. Should we attempt to compete on such projects with bigger countries, we will invest precious manpower and talent in endeavors where we stand little chance of making independent progress, and where what we might achieve will not constitute a scientific achievement. At the same time, bigger countries that can venture into scientific fields such as rocket science, and into projects that require resources on a massive scale, can delegate smaller though not necessarily less important tasks to a small country which does not compete with them but attempts to find its place in international science.

In sum, the possibilities of developing science in a small country are limited, and the chances of raising geniuses are small. A small country can expect, at most, one major scientific work in one discipline in one generation, and another major scientific work in a different discipline in a different generation. It runs the risk of not rising even to that should it detach itself from the circuit of international science. In that case, its scientific system will become mediocre and stagnant, and soon it will find itself straggling behind, just as things happened in Israel between the 1930s and the 1950s. Our only resort is to integrate with a larger scientific system, try to put its resources to the best possible use, and adapt to its evolutions. At times, marginality can prove advantageous, as it allows one to ignore scientific fads and to work in neglected fields. On the other hand, should we cling to our superiority complex, to our belief in the "chosen people," and compete with bigger countries, we will expend our limited scientific forces and resources in vain.

Note

1. The article "Science in a Small Country" was originally published in *The University: A Journal of the Hebrew University of Jerusalem* 10, no. 2 (1964): 12–17 (in Hebrew). It is presented here, for the first time, in English, in the translation of Anna Grinfeld. The permission of Miriam Ben-David to publish this paper is gratefully acknowledged.

2

Astrology and Science: A Renaissance Problem

Brendan Dooley

From the standpoint of many thinkers in the Renaissance, astrology was science.[1] It comprised a body of knowledge that fit the criteria of verification commonly accepted for confirming information and establishing certitude about the natural world. It derived from authoritative traditions rooted in admired ages and places, illuminated by ancient wisdom. The most respected intellects had set great store by it from time immemorial. It made sense according to prevailing ideas about how the world and human nature worked. Its language was embedded in the very discourse whereby the results of scientific investigations were expressed. It seemed to conform to observations and experiences accumulated over time. Its methods were the methods of all knowledge-gathering. It used an experiential, not an experimental approach; and as such it belonged to Renaissance science and only partly to ours. And to the extent that Renaissance thinkers began to invent modes of knowledge-testing to which it could not conform, it gradually lost its grip on Renaissance minds and was superseded by other approaches.[2]

Thus the story of astrology and science in the Renaissance is largely the story of science in general. No wonder Giovanni Battista Riccioli included both astrologers and astronomers together in the list he compiled at the end of the seventeenth century of experts on celestial matters from ancient to modern. And no wonder that, well into the seventeenth century and beyond, it kept its position among the studies associated with "mixed mathematics" as taught in the medical schools. And the works and days of its practitioners deserve all the attention that, for instance, the *Cambridge History of Early Modern Science*, or indeed the *Cambridge History of Renaissance Philosophy* have devoted to them. To these anthologies we also refer for a fuller discussion of the problematic term "science" in this context, adding only that for the sake of convenience "science" and "natural knowledge" will be used interchangeably in deference to historical usage when referring to our period, and clear indications will be given when anything specifically relating to the notion of "modern science" is in play.

To understand the full implications of the theme of astrology and science, therefore, is to understand a basic episode in the formation of humanity's outlook on nature, and perhaps, to throw more light on the complex pedigree of a category of experience, "science," that since the Renaissance has loomed ever larger in human life. In this exploration, we will take three well-known and deeply contrasting interpreters of natural knowledge to epitomize the relation between astrology and science in the Renaissance. For our purposes, Gerolamo Cardano exemplifies how astrology was bound up with other natural knowledge, Francis Bacon shows how other knowledges could lead to innovations in astrology, and Johannes Kepler shows how astrology could lead to innovations in other branches of science. All three were highly controversial when they wrote and their writings subsequently spawned intense debates among scholars attempting to appraise their contributions. We will mention rather than engaging fully with the scholarship on each one, as we focus on the aspects of their work that elucidate our topic.

Gerolamo Cardano wrote, according to his own last count in 1575, nearly a hundred books on almost every field of human endeavor in his time (*On Water, On Dialectics, On the Studies of Socrates, On Civil Prudence, In Praise of Nero,* are some of the less known), although he is remembered in science history textbooks mainly for contributions to algebra, game theory, and machine construction. Most recently attention has been drawn to his voluminous medical works, which summarized theory and practice in an age to some degree torn between Galen and Hippocrates, and he is now studied for his (not unrelated) astrological works—also because of his having produced the last great commentary on Ptolemy's astrological manual, which he entitled, *Claudii Ptolemaei Pelusiensis libri quatuor, De Astrorum Iudiciis cum expositione Hieronymi Cardani*. Though he referred to himself in print mostly as a "Milanese" in virtue of his origin or as a "medicus," in view of his profession, he would have considered that his breadth of knowledge qualified him as a "philosophus." He attempted not once but twice to conceive of a unified theory of all things.

His massive *De subtilitate* (in twenty-two books), later reworked into *De rerum varietate* (in seventeen books) contains the substance of his thinking about humans within the natural world, and the heavens above, problems which he and his contemporaries regarded as the highest form of natural knowledge. It was not, he said in his autobiographical writings, his greatest work, but likely the one that would be most widely read.[3]

Cardano conceived of *De subtilitate* and its sequel as a kind of encyclopedia of natural knowledge (at one point he calls the later installment a "history of the universe and the world"), organized around the concept of "subtlety." This concept embodied his basic view of the universe as a dense web of occult connections full of arcane meanings forming patterns which humans might endeavor to understand in order to achieve mastery over nature, themselves, and history. "Subtlety," in his definition, referred to "sensible things that are difficult to comprehend by the senses, and intelligible things difficult to comprehend by the intellect." Amplifying this, he says, "it is the science of some things that are and some that only seem to be; some which are seen with us sleeping, others with us awake." Characteristic of his effort to use hidden capacities as well as discover hidden meanings, he asserted the concept for the work came to him in a dream; and the method of compilation appears to give equal space to inspiration and pattern-recognition. Although he intended the work to be systematic, like those of his ancient models Aristotle and Pliny, and he divided it into categories like "1: On beginnings," "2: On the Elements," "3: On the Heavens," "5: On Metallic Things," "11: On the Necessity of Man," "13: On the Senses," "16: On the Sciences" and "17: On the Arts," nevertheless his inclusion of material in each of these categories was subject to the meanderings of his own mind over entire ranges of ancient and modern sources. He provided answers to major questions and minor: why the sea is salty as well as "why asses are stupid."[4] Where he is not satisfied with the accepted schemes of organization he invents his own: the four excellent things in nature, the five kinds of stone, the seven human calamities, the nine kinds of animals.[5]

He avoids a unilateral approach, and endeavors rather to proceed back and forth between universals and particulars. Laws come from experience and return to organize the results of yet other experience. Everyday observations form the basis for weighty conclusions, as when the presence of salt in the sea goes to prove that the world is eternal. The work is as much an essay on a method of inquiry as on the matter at hand. He subjects his hypotheses to tests, and draws more hypotheses from his tests. Sometimes he thinks the best answer is no answer. Gold should taste better than silver because it is higher on the scale of excellence; however, it has no taste at all. He also uses particular and sometimes rather homely tests to demonstrate this or that much more weighty point. To explain the effect of tides on the surface of water he suggests pinching the skin to form a "tumor" and observing what this does to the surrounding skin. The approach seemed to engage enough readers for numerous reprintings, although it also gained him the ridicule of Julius Scaliger—damagingly or innocuously, according to different modern interpretations.

The main force in the universe is celestial heat, which causes motion and change by action in varying degrees on the living souls existing in all matter, including animals and humans. It seems to take the place of the peripatetic notion of substantial form and introduce a hylozoic position closer to Plato, the Neoplatonists and even Averroes—whose ideas about unity to some degree inspired his own, expounded in a particular treatise *De uno*. In a posthumously published treatise on nature, he remarks, "whoever has written about the soul says it is either the author of nature, or else is nature itself."[6] At one point in *De subtilitate*, he says natural heat and the soul are also the same thing.[7] The sun and the other heavenly bodies possess this heat and communicate a generative force to the world. This is evidently the basis of astrology. Indeed, "no one may deny" that such forces exist, he said; because "they rule all mortal things," in spite of any assertions to the contrary by "the ambitious and the impious."[8]

Cardano's scheme of natural philosophy challenged the astrological tradition as described by Ptolemy, by removing fire from the list of elements, because of its special role in the universe, leaving only water, air, and earth, so there could be no four-part division of the constellations according to their relation to the elements; and he overturned the division of the planets according to their relation to the four qualities by removing cold and dry and leaving only hot and moist. Moreover, he was able to demonstrate that the *Centiloquium*, widely used by practitioners and long thought to be an astrological work by Ptolemy, was actually a later forgery. In his commentary on the *Tetrabiblos*, he attempted to restore ancient astrology to its former splendor. He identified symptoms, in his source, of the earlier, and in his mind unreliable, Chaldean and Egyptian traditions, so these could be expunged. Instead, he would place before the public a purified version, fit to silence the skeptics and restore the prestige of current practice. He expanded upon Ptolemy's comments regarding where astrology was more reliable (say, in dealing with the weather) or less (say, in predicting the exact course of a person's life). He drew out to greater length (in passages later carefully read by Kepler) Ptolemy's conjectures concerning the rapport between the chief aspects and the harmonies in music. Where Ptolemy had merely mentioned the significance of comets, he added a whole section, based on the wide-ranging literature on the topic in his own time, by Regiomontanus, Pontanus, and Nifo. To the astrologer's toolkit, he confirmed the modern addition of conjunction theory in relation to the fates of nations, drawing upon medieval, and especially Islamic, traditions subsequent to Ptolemy. And as part of his set of demonstrations of astrological technique, he introduced and explicated the geniture of Jesus Christ, answering definitively, so he hoped, the questions raised by a debate that had begun practically with the Renaissance. If astrology could even illuminate theology, the queen of the sciences, its value was secured.

Bacon belonged to the next generation (d. 1626); and as one of the paradigmatic intellectual innovators of the Renaissance, he considered a purely mathematical approach to celestial matters to be highly defective.[9] "Astronomy, as it now stands, loses its dignity by being reckoned among the mathematical arts, for it ought in justice to make the most noble part of physics."[10] He agreed with the division of tasks between astrology and astronomy, with the first comprising knowledge about the significance of the movements, and the second about the movements themselves; but he considered the two approaches to be worthy of study together. He thought astrological knowledge was real, but its use and practice needed to be reformed. "Astrology is . . . so full of superstition that scarce anything can be discovered in it, though we judge it should rather be purged than absolutely rejected." The basic principle of stellar influences upon which astrology was based he found to be unimpeachable, and likewise the notion of a rapport between the microcosm of human affairs and the macrocosm of celestial ones. He said, whoever understood the connection "between terrestrial and celestial things, and well understands the more general appetites and passions of matter, which are powerful in both, may receive a clear information of what happens above from that which happens below; and from what passes in the heavens, he may become acquainted with some inferior motions hitherto undiscovered, not as these are governed by those, but as they both have the same common passions."[11]

Stellar forces were not only heat and light, in Bacon's view, but included "certain other influences," which varied in kind and strength according to the planet or star in question, and according to the relation between the planet or star and the earth. He did not venture to pronounce on the precise character of these other influences; it was enough to know they existed and worked observable effects in the world, and to understand something of the dynamics. He outlined the studies that might be done in this connection. Mixtures of rays (conjunctions, oppositions, and other aspects) were to be taken into account, as well as the passage of the wandering stars through the zodiac. Angles were fundamental, and so were apogees and perigees, acceleration and retardation. The particular essence of a star was important—whether fixed or wandering, whether one color or another, whether twinkling or not, large or small. Tradition could be a help where observation was lacking, in terms of the ages-old agreed-upon characters of the stars (the beneficence of Jupiter and Venus, the maleficence of Saturn and Mars). When all of these features were calibrated and described, the result, in Bacon's view, would be a "sane astrology," founded upon the same "physical reasons" as a "living astronomy."

The trouble with astrology, according to Bacon, was not the principles per se, but the astrologers and their traditional practices for judging of the future. In other words, events and characters were the products of a network of cosmic connections governing the great body of the world. However, "we reject, as an idle figment, the doctrine of horoscopes, and the distribution of the houses, though these are the darling inventions of astrology, which have kept revel, as it were, in the heavens."[12] Instead, in the spirit of the Great Instauration, of which *The Advancement of Learning* formed a part, that is, Bacon's project for a "total reconstruction of sciences, arts and all human knowledge, raised upon the proper foundation,"[13] he would reform astrology by expunging the uncertain and the patently false and collecting what could be verified. There were, in his mind, four routes to truth in this science. The first two were by "experiments" past and future, by which he meant experiences of predictions gone right; next came traditions, and finally "physical reasons." He proposed a collaborative effort of hitherto unimagined proportions, to proceed from the mass of evidence to the formulation of new laws. "The astrologers may, if they please, draw from real history all greater accidents, as inundations, plagues, wars, seditions, deaths of kings, etc., as also the motions of the celestial bodies . . . to . . . erect a probable rule of prediction." Such information was of course to be carefully scrutinized. "All traditions should be well-sifted, and those thrown out that manifestly clash with physical reasons, leaving such in their full force as comport well therewith." This, he avowed, "we take for the surest guide to astrology."[14]

If such thoughts did not exactly fit the picture, propounded in the Enlightenment, of Baconianism as the incubator of modernity, they nonetheless make perfect sense in the context of the early modern culture we are examining. No modern edition of *The Advancement of Learning* can afford to simply leave them out entirely, as Basil Montagu did in the early nineteenth century; nor can any manual on history of science pass over them in silence as was the practice of George Sarton and other early twentieth-century writers. In fact, Bacon's diffidence with regard to contemporary astrological practice, combined with his acceptance of the principle of stellar influences, fit a pattern shared among other figures who have been considered to be leading innovators in a period of rapid intellectual change.

Kepler contrasted sharply with Bacon, ten years his senior, in terms of method as well as intellectual proclivities.[15] He tended toward a priori explanations, which he attempted to defend first on a priori grounds (elegance, aesthetic, mystical) and only then by empirical data. Number mysticism, drawn from Platonic neopythagoreanism, informed the world picture he first presented in the *Mysterium Cosmographicum*, where the heliocentric model, derived from Copernicus, was based on the existence of six planets and six geometrical figures including the five regular solids nested one inside the other, plus the outermost sphere. Plato, in those pages of the *Timaeus* which contained so much of the natural knowledge he ever transmitted, had coordinated four of the solids with the four elements (cube: earth; tetrahedron: fire; octahedron: air; icosahedron: water). Kepler ordered all five by importance from a mathematical viewpoint (cube, tetrahedron, dodecahedron, followed by octahedron and icosahedron), and related them to the planets, with the three most important ones including the earth and outward toward the fixed stars, and the two remaining ones toward the sun. He found a peculiar confirmation for this arrangement in the relation, explained by Plato, between the study of harmony and the five regular solids, pointing out that just as the solids were ordered in a hierarchy, there were three chief intervals (octave, fourth, fifth) and two subordinate ones (third and sixth). His calculation of hypothetical distances of each planet from the sun on the basis of the related

Figure 2.1

Kepler's Polyhedric Model. *Opera Omnia*, ed. Ch. Frisch (Frankfurt: Heyder u. Zimmer, 1858), p. iv.

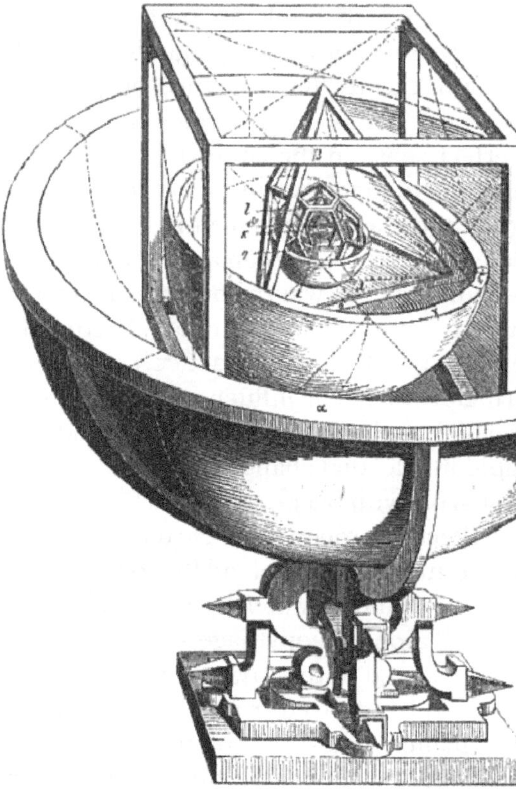

α Sphaera Saturni. β Cubus. γ Sphaera Jovis. δ Tetraëdron. ε Sphaera Martis. ζ Dodecaëdron. η Orbis Terrae. ϑ Ikosaëdron. ι Sphaera Veneris. κ Octaëdron. λ Sphaera Mercurii. μ Sol, Medium sive centrum immobile. (Comp. Fol. 214.)

polyhedron seemed to coordinate with the latest planetary distances calculated by Michael Maestlin, his mentor.

The question for Kepler was not whether planetary influences existed or could be measured, but whether the well-known characteristics agreed upon by generations of astrologers could be predicted by his geometrical model. The results were striking. Mercury's famed inducements to celerity referred to its octahedron, which could be turned cleanly on two angles, whereas the other figures, in his view, could not. Jupiter and Venus were supposed to indicate cause and effect, or fertility and offspring: and indeed, the pyramid (tetrahedron) of Jupiter, presumably, impregnated, whereas the icosahedron of Venus received the impregnation. In addition, "the female is always various and mutable," he said, and this was confirmed by the polyhedron of Venus. Jupiter's beneficence was noted already by Ptolemy. Kepler saw the same, but transposed to the heliocentric universe; and he found an explanation once again in its related pyramid. Although generally beneficent, Jupiter was hostile to the other two superior planets, Saturn and Mars, a fact apparent in their three solids: the cube, the tetrahedron, and the dodecahedron:

> Also among their three solids absolutely none of their observable properties agrees, though Mars conspires with Saturn in malice alone. To this I relate the variability of their angles, which is peculiar to them, and common to both.

The beneficent planets could also be explained by their solid geometry: "the constancy of the angles between their edges alone, is evidence of benignity."[16]

In 1606, he finished his key work on the orbit of Mars, the *Astronomia nova,* where he announced his first two laws of planetary motion, stating equal areas and elliptical orbits, the latter of which posed serious difficulties for his earlier concentric spheroid model involving the polyhedrons. In this work, he tended more toward an empirical approach than in his previous productions, collecting and interpreting the observations of Tycho Brahe and Maestlin, modifying his model in tandem with what he saw. The reason for the form of the orbits still eluded him, but he liked the neoplatonic concept of an *anima mundi* or world soul capable of making judgments. The planets, he suggested, have a "mind," whereby they steer this way and that, always according to a pattern—in this case, a non-circular one. The new discoveries made no contribution to astrology, he said; although everyone knew (and he clearly implied an autobiographical reflection here) the chief reason for doing astronomy was to further astrological interests.[17]

In his later work, Kepler kept the polyhedrons in his cosmology, but used them less dogmatically and more heuristically, also in view of the new discoveries. A new idea began more and more to take hold of his imagination: the relation between planetary calculations and the harmonic series. Hints of this idea, elaborated from a theory of Ptolemy, already appeared in the *Mysterium,* but the mature version emerged in the *Harmonice Mundi* of 1618. He found a striking correspondence between the series and the planets' angular velocities at perihelion and aphelion, and calculated the harmonies of all of them beginning with Saturn at aphelion, taken to be the lowest note, "G." That the chromatic scale had twelve tones he found to be particularly significant in light of the twelve houses of the zodiac. Here too he elaborated on the cause of motion in the universe, based on the *anima mundi* or world soul, drawing upon the animal analogy used by Giordano Bruno and other neoplatonists, an approach that led him far away from geometry and mathematics, and toward a general philosophy of matter. "Just as other animate beings consume food and drink, so the Earth also must take some kind of material from definite channels, to brew from it such a multiplicity of substances, because nothing is made from nothing."[18] Indeed, "as the bladder pours out urine, so the mountains pour out rivers; as the body produces excrement of sulphurous odor and farts which can even be set on fire, so the Earth produces sulphur, subterranean fires, thunder and lightning."[19] And so forth. As to the type of soul the Earth had, "it seems plainly to be a sort of flame."[20]

Within this view of nature there was plenty of room for astrology, and Kepler applied what he knew in various prognostications and in the composition of nativities for the court, as part of his employment by Emperor Rudolph II. He never doubted that the planetary aspects powerfully influenced the world ("for I saw with great consistency that the state of the atmosphere was disturbed whenever planets were either in conjunction or configured in the aspects commonly spoken of by the astrologers"[21]) or that this influence was transmitted to the *anima mundi.* Individual souls responded, much as the Earth itself, to the stellar influences, and the collective impact of these effects could influence mass actions: hence the coincidence of wars and other major human events with the major conjunctions. Furthermore, souls received at birth an impression of the zodiac, which created dispositions of character. "For since the vital faculty, lit in the heart, and burning as long as life persists, is a kind of zodiac circle, since its essence consists in activity, and in a flow of flame, as it were, the result is that the whole sensible shape of the zodiac flows into it when it has been freshly lit at birth."[22] The planets, the ascendent house, the descendent 180° away, the mid-heaven or cusp of the tenth house, all sent particular harmonies to the soul, forming predispositions, inclinations, and shades of character. The study of a nativity was a key element in judging and understanding a human being. For instance, he hypothesized, "those who are born at a time of many aspects among the planets generally turn out hard-working and industrious."[23] His own nativity he adduced as an example.

Also in the *Harmonice*, Kepler engaged in the nearly obligatory practice of denouncing the naïve strain of working astrology, as he had done already in his annual predictions of 1598 and 1599. Rereading Giovanni Pico della Mirandola's by now classic diatribe, he said he was led "to confirm my condemnation of a great many superstitions,"[24] and attempted to find ways of restoring the conceptual foundations. Experience showed no presumed celestial effects to be ineluctable, and no planetary dispositions to be determining. The space of predictions obviously had to be redefined and circumscribed. Too often, he said, the planets were regarded as omnipotent gods, inexorable and capable of creating effects at will, although few observers of those effects had considered exactly what might be the physical causes, that is, the actual mechanisms that brought them about from the standpoint of current natural knowledge. The polyhedron theory was one way to do this, but this was only one portion of a multi-level project. In fact, he began the work of reforming astrology in 1602 with the publication of his *De fundamentis astrologiae certioribus*. True to its title, there he proposed to demote the zodiac from its primary role, for the same reasons that Pico had poured scorn on it: because the precession of equinoxes across the centuries had put the actual positions of the stars out of phase with the positions supposed in the traditional view. He would consider the names of the houses to be irrelevant and put much more emphasis on the Ptolemaic aspects, to which he added three new ones: the quintile (72°), bi-quintile (144°), and sesqui-quadrate (135°). He would also follow the contemporary trend, influenced by Arab astrology, in giving greater attention to conjunctions.

Kepler thought he saw his views spectacularly vindicated by the appearance of a new star in the constellation Serpentarius in 1604, close to where a Great Conjunction of Jupiter and Saturn had occurred the year before in Sagittarius, signaling the beginning of the Fiery Trigon or triplicity of Sagittarius, Aries, and Leo, whose slowly gyrating 30° of arc it would continue to traverse for the next 200 years. Such events, he was convinced, were obviously significant—although interpreting the significance was (he opined) probably beyond the abilities of any of the observers. In one work dedicated to the phenomenon, he insisted, "The star's significance is a difficult matter to establish and we can be sure of only one thing: that either the star signifies nothing at all for Mankind or it signifies something of such exalted importance that it is beyond the grasp and understanding of any man."[25] In his definitive work on the topic, the *De stella nova*, published in 1606, he indicated his hope that the event signaled a "beautiful and lasting ending" to human suffering, but he reiterated his doubts about humans' ability to interpret the signs in any way that might be useful to them.[26] The best option for humankind, in his view, was to take the occasion as a prompting to turn more piously to the things of faith, and repent before it was too late.

Later in the *Tertius interveniens* of 1610, he assumed the role of a mediator between the naïve astrologer and the skeptic, with a "warning, to true Theologians, Physicians and Philosophers . . . when simply rejecting the stargazing superstitions, not to throw the baby out with the bathwater and thereby ignorantly act in contravention of their Professions."[27] The study of astrology, he pointed out, was no more uncertain than, say, the study of simples in medicine: both were arts of practice as much as bodies of knowledge, and both relied on the skill of the practitioner in applying the knowledge to the case. Moreover, "in the beginning there are also many failed experiments."[28] Any attempt to forbid even the silliest forms of astrology, such as the various imperial injunctions in the ancient world and the recent Bull of Sixtus V, would cancel an important road to physical and theological truth, and made no more sense than forbidding the study of anatomy. After all, he knew well what recent scholars have now discovered, that without astrology, he himself might not have been drawn in the direction of his major accomplishments.

For each of these three figures, astrology was at once more and less than science. It was more because it involved a particular outlook on the world as well as a way of doing knowledge. It was less than science because its practitioners in all ages, no differently from Ptolemy himself, believed that many of its propositions had not yet been verified, and had simply been received on the strength of the tradition and their evident plausibility. Ptolemy asked:

> If, then, a man knows accurately the movements of all the stars, the sun, and the moon, so that neither the place nor the time of any of their configurations escapes his notice, and if he has distinguished in general their natures as the result of previous continued study, even though he may discern, not their essential, but only their potentially effective qualities, such as the sun's heating and the

> moon's moistening, and so on with the rest; and if he is capable of determining in view of all these data, both scientifically and by successful conjecture, the distinctive mark of quality resulting from the combination of all the factors, what is to prevent him from being able to tell on each given occasion the characteristics of the air from the relations of the phenomena at the time, for instance, that it will be warmer or wetter? Why can he not, too, with respect to an individual man, perceive the general quality of his temperament from the ambient at the time of his birth, as for instance that he is such and such in body and such and such in soul, and predict occasional events, by use of the fact that such and such an ambient is attuned to such and such a temperament and is favourable to prosperity, while another is not so attuned and conduces to injury? (*Tetrabiblos* I,2)

In Ptolemy's view, as in the view of the Renaissance thinkers so far analyzed, and others besides, astrology was a faith as much as a science: a faith that eventually better data-gathering would make real astrological knowledge possible, so that this knowledge could be used effectively to improve human life. In the absence of any single conceivable decisive proof, the value of astrology would be attested by the continuing commitment of its admirers and the accumulating evidence of its successes, which progress guaranteed would be more impressive in the future than in the past.

In an emerging world where celestial motion was to be explained by other means, and the claims of science were to be reduced to what could be verified by tests reproducible in a laboratory, and the mysteries of the cosmos no longer demanded an immediate explanation, astrology came to be pushed to the outer edges of the intellectual world, into the area occupied by theology and other speculative arts. And when Louis de Jaucourt, in his article on "Influences" in the French *Encyclopédie*, labeled astrological knowledge "pretendu" and submitted it to a withering critique, there was no one left to defend it. The spell it had long cast over European minds had finally been broken.

Notes

1. So was magic. Here I side with Cesare Vasoli in his Introduction to *Magia e scienza nella civiltà umanistica* (Bologna: Il Mulino, 1976). In particular relation to astrology and science, see the special issue, edited by Lauren Kassell, of *Studies in History and Philosophy of Biological and Biomedical Sciences,* vol. 41 (2010). In addition, Eugenio Garin, *Astrology in the Renaissance: The Zodiac of Life* (Boston, MA: Routledge, 1983, Italian original Laterza 1976); Patrick Curry, ed., *Astrology Science and Society: Historical Essays* (Wolfeboro, NH: Boydell Press, 1987); Idem, *Prophecy and Power: Astrology in Early Modern England* (Princeton, NJ: Princeton University

Press, 1989); Paola Zambelli, ed., *"Astrologi hallucinati": Stars and the End of the World in Luther's Time* (New York: Walter de Gruyter, 1986); Idem, *The "Speculum Astronomiae" and Its Enigma: Astrology, Theology, and Science in Albertus Magnus and his Contemporaries* (Boston, MA: Kluwer Academic, 1992); Idem, *L'apprendista stregone: astrologia, cabala e arte lulliana in Pico della Mirandola e seguaci* (Venice: Marsilio, 1995); articles in *Horoscopes and Public Spheres*, ed. Günther Oestmann, H. Darrel Rutkin, and Kocku von Stuckrad (Berlin and New York: Walter de Gruyter, 2005).

2. Concerning the waning of astrology note H. Darrell Rutkin, "Astrology," in *The Cambridge History of Science*, vol. 3: *Early Modern Science*, ed. Katherine Park and Lorraine Daston (Cambridge: Cambridge University Press, 2008), 542–63.
3. All citations are from the *Opera Omnia,* 10 vols. (Lyons: Jean Antoine Huguetan/Marc Antione Ravaud, 1663). Here, vol. 1, p. 70, from his *De Libris Propriis.*
4. *Opera Omnia*, vol. 3, book 10.
5. On these features see Ian Maclean "The Interpretation of Natural Signs: Cardano's De Subtilitate Versus Scaliger's Exercitationes," in *Occult and Scientific Mentalities in the Renaissance*, ed. Brian Vickers (Cambridge: Cambridge University Press, 1984), 242.
6. *De Natura* (op. vol. ii, chap. 3).
7. *De subtilitate*, lib. 2, de elementis.
8. *De rerum varietate*, II: xiii.
9. Concerning Bacon and his project, Stephen Gaukroger, *Francis Bacon and the Transformation of Early-Modern Philosophy* (Cambridge and New York: Cambridge University Press, 2001). Concerning the impact, consider Nicholas Popper, "'Abraham, Planter of Mathematics': Histories of Mathematics in Early Modern Europe," *Journal of the History of Ideas* 67 (2006), 87–106.
10. Francis Bacon, *Advancement of Learning*, vol. 3 of *The Works of Francis Bacon*, ed. James Spedding and Douglas Denon Heath, 14 vols. (London: 1857–74), book III, chap iv.
11. *Advancement of learning*, book III, chap iv.
12. Advancement of Learning, Proemium.
13. *Advancement of learning*, book III, chap iv.
14. *Advancement of Learning*, ed. Edward Creighton (New York: Colonial Press, 1900), III: 4 (p. 90 in this edition). Not in the least Whiggish of course is Paolo Rossi, *Francesco Bacone: Dalla magia alla scienza*, 2nd edn. (Turin: Einaudi, 1974); and especially see D. P. Walker, "Spirits in Francis Bacon," in *Francis Bacon: Terminologia e fortuna nel secolo XVII* (Rome: Edizioni dell'Ateneo, 1984), 315–27. Of course, I am not confusing Bacon's empiricism with the "empirics" whom he condemns. See Didier Deleule, "Experientia-experimentum chez F. Bacon," in Ibid., 59–72.
15. Concerning Kepler, consider Sheila J. Rabin "Kepler's Attitude Toward Pico and the Anti-Astrology Polemic," *Renaissance Quarterly* 50 (1997): 750–70; as well as Gérard Simon, *Kepler: Astronome, Astrologue* (Paris: Gallimard, 1979), there is Fernand Hallyn, *Structure poétique du monde: Copernic, Kepler* (Paris: Seuil, 1987) and J. V. Field, "A Lutheran Astrologer: Johannes Kepler," *Archive for the History of the Exact Sciences* 31 (1984): 189–272,

all to be compared with Bruce Stephenson, *Kepler's Physical Astronomy* (New York: Springer-Verlag, 1987).

16. Modified from Kepler, *The Secret of the Universe,* tran. A. M. Duncan (New York: Abaris, 1981), 115–6. In addition, here, Mary Ellen Bowden, *The Scientific Revolution in Astrology: The English Reformers, 1558–1686,* PhD dissertation, Yale University, 1974, 112.
17. John North, *The Fontana History of Astronomy and Cosmology* (London: Fontana Press, 1994), 309–26.
18. Kepler, *The Harmony of the World,* tr. and notes by E. J. Aiton, A. M. Duncan, and J. V. Field (Philadelphia, 1997), 363.
19. Ibid., 364.
20. Ibid., 366.
21. Ibid., 360.
22. Ibid., 374.
23. Ibid., 375.
24. Ibid., 360.
25. *Opera Omnia,* ed. Christian Frisch, 8 vols. (Frankfurt: 1858–71), vol. 1, *Bericht vom neuen Stern, 1604,* 396—M. A. Granada, "The Discussion between Kepler and Roeslin on the Nova of 1604," in M. Turatto, S. Benetti, L. Zampieri, and W. Shea, ed., *1604–2004: Supernovae as Cosmological Lighthouses,* ASP Conference Series, vol. 342 (Provo, UT, 2005), 37.
26. "The Discussion between Kepler and Roeslin," Granada, 41.
27. *Tertius Interveniens,* title page, in *Opera Omnia,* vol. 1.
28. *Opera Omnia,* vol. 1, 630, thesis 111.

3

Social Framework of Early Theoretical Science

Dmitri Panchenko

Theoretical science seems to have independently emerged only once in the course of history. This happened in the sixth century BC, within the Greek world, in Ionia, in Miletus. Wherever people do theoretical science, they follow the path of Ionian Greeks. This is not of course to say that the Greeks were the first to have displayed an interest in studying celestial phenomena or geometrical figures. A kind of scientific research was quite developed in Egypt and especially in Mesopotamia. The most distinctive feature of Greek science, as compared with that of Mesopotamia or Egypt, pertains to the role of hypotheses and proofs. No scientific conjecture or demonstration is known from sufficiently numerous Near Eastern texts. Again, it is not likely at all that no hypothesis was ever formulated in Babylonia, Assyria, or Egypt. But such sparse hypotheses did not constitute there a tradition of argumentative knowledge, of improving and advancing knowledge through criticism and demonstration.[1]

It was very different in Ionia. Already Thales, the founder of both theoretical science and philosophy of nature, advanced an elaborate scientific theory—a theory of solar eclipses. The Greek word for "eclipse" means "disappearance," but Thales asserted that at a solar eclipse the sun was screened from us by the moon. This was essentially correct. Now, Thales was a citizen of Miletus, a city-state that founded many dozens of settlements from the Black Sea coast to the Nile Delta and that maintained regular connections with most of them. It was therefore known in Miletus that the same solar eclipse is experienced as total in one area and seen as partial in adjoining regions, and not seen at all elsewhere. In Thales' theory, this meant that the moon was smaller than the sun. But that solar eclipse which Thales happened to predict (not without a degree of good luck), the eclipse of May 28, 585 BC, appeared as total within a band of more than two hundred and fifty kilometers. It followed, then, that the moon was at least as large as that and the sun yet much larger. Although everybody knew that objects appeared smaller as distance increases, the consequences of Thales' theory were striking. In order to meet criticism and prove that he was right, Thales determined the ratios between the apparent sizes of the sun and moon and their daily orbits. He found, quite near to the truth, that both ratios were about 720 to 1. Skeptics had to admit that, if the sun and moon were about one foot in length, their daily courses would be less than one thousand feet, which was patently absurd. Thus Thales not only replaced divine command with natural processes, but also with amazing boldness suggested a picture of the sky in which incredibly huge celestial bodies rotated with unbelievable speed.[2]

Thales left no book, and we do not know whether he dared to draw the outlines of cosmic edifice. But we know for sure that his immediate follower, Anaximander, did so. Anaximander was the first man on earth to publish a theoretical treatise (soon after 547 BC). Wind and rain, lightning and thunder, earthquakes and eclipses were explained in this treatise as natural phenomena. Anaximander approached in the same way the origin of animal life and man as well as the formation of celestial bodies, the earth and the sea. His treatise has been lost and therefore details of his cosmic edifice remain problematic. The core idea is nevertheless clear. According to Anaximander's revolutionary view, the earth is suspended and has no support. It does not fall down because of symmetry and indifference. It occupies the centre of the cosmic order. What we see above will be seen below, from the opposite side of the earth in half of a day. It is true that heavy bodies move down, but what is down and what is up for the earth so situated? The earth has no preferred direction to fall in and therefore remains stable. The geocentric system of the universe was thus introduced, which efficiently worked for two thousand years.[3]

Now Anaximander's brilliant theory of an earth hanging in space without support meant polemics against the view of his predecessor and teacher, for according to Thales, the earth was supported by water. We are not informed what precisely Thales had in mind. Perhaps, he thought of a boundless expanse of water—but water normally needs a container. Perhaps, he imagined such a container—but what would support such? Anaximander tried to suggest something better. He also deviated from Thales' view concerning the nature of eclipses.

The evidence that we have on this point is not unambiguous, and yet we may safely infer that Anaximander's suggestion was a retrograde step since Thales' view was correct. On the other hand, Anaximander accepted and developed Thales' conviction that the sun and the moon are immensely greater than they appear to the senses. Anaximander asserted that the sun was no smaller than the earth.

The third in the series of great Milesians, Anaximenes, rejected the best of Anaximander's theories and reintroduced a support for the earth. I cannot share a standard scholarly opinion that Anaximenes was simply not bold enough. Only intellectually boldest people could have been engaged in theoretical research when this was an entirely new activity. Anaximenes must have had a reason. He probably became aware of the phenomenon of polar day. The fact that there are places on earth where the sun does not set seemed to have made impossible Anaximander's theory in which the sun passed below the earth as well as above it. Whatever Anaximenes' reason, we see another polemical approach and another new cosmic system. The rising and setting of luminaries were taken to have been an optical illusion. A broad essentially flat earth was kept on a cushion of air. That was not just a return to an idea of Thales. Anaximenes substituted air for water, and this was a brilliant solution. Invisible, but proven to exist and resist pressure in an inflated skin, Anaximenes' air, unlike Thales' water, needed neither container nor support. Anaximenes' astronomical system was influential just for a short time. The data were assembled to show it was wrong. Parmenides, a younger contemporary of Anaximenes, was the first to advance the theory of a spherical earth.[4] Not without resistance, which was rather a fruitful examination, his theory was finally firmly established in Greek science. The same happened to Thales' theory of solar eclipses, now supplemented by an elaborate theory of lunar light (Panchenko 2002a).

Early Greek astronomy also addressed the problems related to the variation of the length of day in accordance with both seasons and different locations. It meticulously studied the relation of the course of the sun (seen only during the day) with tracks of the stars (seen only during the night). The yearly course of the sun along the ecliptic, or along the belt of the Zodiac, was firmly established by Oinopides of Chios (around the middle of the fifth century BC). Oinopides also wrote the first introduction to the study of astronomy in which he combined his knowledge of the movement of celestial bodies with his expertise in geometry.

Elaborate argumentative geometry is another glory of early Greek science. Thales, its founder, was the first to prove general geometrical propositions.[5] Pythagoras proved the theorem that bears his name.[6] Hippasus proved the incommensurability of a square's diagonal with its side; he thus truly inaugurated theoretical geometry since there was no way to ascertain the correctness of his proposition through observation and measurement. A contemporary and compatriot of Oinopides, Hippocratus (not to be confused with his more famous contemporary of Cos), wrote the first introduction to the study of geometry, a predecessor work of Euclid's *Elements*.

A similar striking and rapid development took place also in the realm of natural philosophy. I refer to the Presocratic theory of matter. Thales, proceeding from the assumption that nothing comes out of nothing, suggested the idea of basic unity of all things.[7] They all emerge from a common eternal source through the process of transformation and they also resolve into this source. He identified such a source with water. His successors shared his assumption, but all of them tried to improve upon Thales' concrete proposal. Some of them, like Parmenides, emphasized the logical difficulties related to the ideas of the emergence, of coming to be, of change. One of the fruits of this polemics was the atomistic theory.[8]

We are dealing thus with a truly extraordinary phenomenon in the intellectual history. We are facing something that really calls for explanation.

The problem of what caused Greek unique achievement was repeatedly addressed. The majority of scholars would approach it in terms of the political dimension of Greek life, citing the polis and democracy. Fine observations and ideas were formulated, for instance, by Jean-Pierre Vernant (2004/1962) and especially by G. E. R. Lloyd (1979).[9] But all that is of limited help. The experience of argumentative debate acquired at the assembly or the court could be useful, of course, for debating scientific matters. However, the strategy of legal or political debate differs much from that of scientific inquiry, as it aims at convincing listeners in a particular moment rather than establishing an "eternal" truth. Furthermore, the polis and democracy as such have nothing to do with the interest in astronomy or mathematics.

Some scholars refer to the encounter of divergent truths, particularly Greek and non-Greek (e.g., Kurt von Fritz, Karl Popper, B. L. Van der Waerden). This explanation has, no doubt, heuristic value and may be relevant to the interpretation of the emergence of philosophical cosmogony. Yet, the history of ideas shows that conflicting truths frequently supported skeptic and agnostic conclusions. In the epoch in question, moreover, not only the Greeks, but also other peoples (including the Assyrians, Babylonians, and Egyptians), had experience of hearing conflicting accounts of gods, cosmogony, and institutions.

Alexander Zaicev (1993/1985),[10] a scholar who dared to approach the so-called Greek miracle with a comprehensive explanation rather than separate observations and insights, focuses on the mechanism for releasing the brakes on the creative potential and for creating a social climate in which every success of intellectual activity receives encouragement independently of its practical applicability, that is, in which the discovery of a new truth could bring fame.

However, while the emphasis on "brake release" and intellectual stimulation can account for the rise of creative activity, it cannot account for the ways in which that activity is realized. First, it is not yet clear how the demonstration of a geometrical proposition or the formulation of an astronomical theory could bring fame (the general public is typically not in a position to evaluate such things, nor is it typically interested in them). Further, the problem remains why curiosity, once encouraged, should necessarily lead to the non-contradictory, systematic, and advancing knowledge. Finally, the efforts themselves, no matter how encouraged, would fade if the adequate form of their application is not found. In other words, there remains the question of the mechanism that organizes the theoretical knowledge.

Such a mechanism seems to have been named by Karl Popper, who emphasized the role of critical discussion and the resulting critical tradition in the amazing development of early Greek philosophy and science:

> The early history of Greek philosophy, especially the history from Thales to Plato, is a splendid story. It is almost too good to be true. In every generation we find at least one new philosophy, one new cosmology of staggering originality and depth. How was this possible? Of course one cannot explain originality and genius. But one can try to throw some light on them. What was the secret of the ancients? I suggest that it was a *tradition—tradition of critical discussion* (Meier and Petersen 1998, 20).

To explain what made the critical tradition possible, however, Popper simply assumed a happy temper of Thales, who encouraged

Anaximander to criticize his own ideas. But one cannot accept his reference to Thales' ability to tolerate criticism as an explanation of a practice that lasted for many generations.

Now, when we speak of critical discussion, we mean by "critical" something more than one's readiness to doubt the correctness of one's assertion. We rather mean a procedure employing factual or logical examination of one's assertion. To conduct critical discussion means to exchange arguments. Critical discussion deals with rationally arguable statements. A rationally arguable statement comes to anticipate questions, such as why should we believe in what you are saying? Consequently, a particular form of human interaction is impressed into the very logic of critical discussion.

And the corresponding pattern of interaction is natural for the upper stratum of Greek society with its free and equal individuals. If Thales, or Anaximander, etc. claimed the recognition of their peers, they had to answer the question of why their unusual assertions were to be met with approval. A demand for a rationally arguable, responsible, statement pertains, however, to a specific set of problems. There is no need for proof or argumentation when one can just go and see or measure. Such a need arises only when the subject is unobservable or (in the case of mathematics) that which cannot be measured. Thus theoretical knowledge emerged as a set of rationally arguable statements about matters that could not be directly observed (or measured).

We may plausibly suppose that the interest in such matters and the interactional pattern characteristic for free and equal individuals were brought together within small associations—both formal and, more typically, informal. Such associations emerged on the basis of common intellectual interests, friendly connections, or perhaps around a distinctively bright person. The associations thus formed were far from suppressing the initiative and aspirations of their members, for whom, on the contrary, the prospect of gaining distinction among their fellows was one of the motives for participation in the association. The fellows, therefore, provided each other both encouragement and critical control. If the demand for responsible statements became a rule within such associations, it was not because the participants in intellectual communication had inherited the pattern of argumentative debate from the court or the assembly. The fact is that they found themselves in a similar interactional situation in the court, in the assembly, and in a discussion of the nature of things: they interacted as free and relatively equal individuals. For the same reason, they were not in need to enjoy the freedom of criticism with a kind permission of the master.

In order to account for the emergence of critical tradition, we should take into consideration the fact that Thales, Anaximander, Parmenides and, in general, most of the Presocratics were doing both philosophy and science. But there is no science without transmission of some technical knowledge. And this implies both tradition and school-like relationship between a teacher and a disciple. We have no reliable information about details of teacher–disciple relationship among the Presocratics, but we may be sure that a former disciple was financially independent from the teacher and school. Only politeness could prevent him from criticizing the teacher if he wished to, and nothing could prevent him from replacing teacher's ideas with his own if he felt that they were superior. Whatever Thales' or anybody else's temper, we may safely eliminate teacher's encouragement of criticism from the account. The encouragement came from elsewhere. I mean the agonal spirit, so characteristic for the Greeks. The desire to surpass one's teacher and predecessors was an efficient driving force behind the critical tradition. It invited improving upon what had been done—and thereby also helped to maintain a certain continuity since the idea of a superior achievement implies a comparison with a previous one.

To sum up, critical discussion and the demand for arguable (or responsible) statements were secured by the interactional pattern characteristic for the small associations based on common intellectual interests. A distinguished member of such an association could become one day a teacher. Critical tradition was maintained, on the one hand, due to the transmission of special knowledge from a teacher to disciples and, on the other hand, due to subsequent attempts of some disciples to surpass the achievements of a teacher.

Yet we cannot explain the emergence and rapid development of theoretical cosmology through certain interactional patterns alone. We must bring into consideration historical circumstances.

Even the earliest scientific theories should have been based on nontrivial data. It can be shown that Thales' theory of solar eclipses depended on a fact established in Assyria and Babylonia in the course of many decades of regular observations, the fact that a solar eclipse can happen only at the time of a new moon. Several generations of observers were necessary to establish such fact firmly. In more general terms, it can be said that Ionian theoretical cosmology was based on the knowledge of Near Eastern, mostly Mesopotamian provenance.

However, the Babylonian and Assyrian observers worked for the temples and the kings. They earned their bread precisely because they had something to say about the will of the gods and the signs given by them through the interpretation of celestial omens. Some individuals might have had their doubts, but the group interest must have effectively prevented the development of thought in the direction of naturalistic conclusions. If some person outside of the group had tried to use the regularity in question while speculating about cosmological issues, his efforts would have been easily dismissed due to the fact that there were competent professionals to form opinions about everything related to celestial phenomena.

The situation in Ionia was quite different. Religious beliefs follow some logic. The gods are supposed to give signs to mighty kings and not to the elected officers of tiny republics. Accordingly, the Greeks had no professional interpreters of celestial omens and were much more open to a naturalistic view of a celestial phenomenon such as a solar eclipse. Furthermore, the absence of professional experts in matters related to heavenly bodies made possible the emergence of the individuals characterized by a new type of intellectual activity. These individuals, the early Greek scientists and philosophers, combined that intellectual curiosity which is natural for a proportion of the population in many societies with some special knowledge, first acquired from abroad and then inherited and enhanced. Because they were free citizens of a free city-state and typically from its upper stratum and because they were not professionals, the philosophers were free to apply their intellectual curiosity to any question they wanted to. Because they possessed some special knowledge (in what later will be called astronomy and mathematics), which the common people did not, their innovative ideas were not to be dismissed easily. And because they were able to demonstrate the power of their knowledge and the advantage of their new approach, they found around them people who shared their intellectual interests. Those small associations, to which I assign an essential role in the development of the critical discussion and critical tradition, could hardly emerge within Babylonian society because there was no room for influential associations of dilettantes near groups of professionals, organized and self-maintained as groups for service and not for "purposeless" theoretical speculations. Although sporadic critical discussion about celestial phenomena was not impossible in Mesopotamia, the conditions for a critical tradition in such matters there were absent.

One more point seems in order. Greek republics were small, but it is significant that the Milesian founders of theoretical science, Thales and Anaximander, came from the leading centre of a prosperous and powerful nation. The emerging theoretical science was about the sun, the moon, and the stars, about the universe and the position of the earth within it. Such a topic was hardly appropriate for a parochial "thinker" from a marginal community: who was he to offer his opinions on such great matters? But by the early sixth century BC, Greek was spoken along many shores of the Mediterranean and the Black Sea, the Greeks maintained contacts with Egypt and Babylon, economy was growing, no serious threat from the outside was yet in sight, and Miletus had a prominent place in this world. Thus, respectable citizens of Miletus were in a relatively favorable position to inaugurate socially acceptable discussion of the structure of the universe and the functioning of its main elements. Thales was truly born in the right place and at the right time. On the one hand, the collapse of the Assyrian kingdom in 612–609 BC must have caused emigration of Assyrian interpreters of celestial omens and thereby dissemination of their astronomical knowledge. On the other hand, there remained a good half a century till the Persian conquest in the 540s for Ionian self-assurance.

So both characteristic features of Greek social life and historical circumstances joined to produce the miracle of early theoretical science.

However amazing, the progress of early Greek science was doomed to be limited. Financially independent, informal, and tiny communities of semi-dilettantes were excellent for producing new fruitful ideas and checking critically previous solutions, but they were not very much conducive to gaining new facts, to enlarging the horizons of knowledge.

Joseph Ben-David defined once the emergence of modern science as "the emergence of a pattern which combines continuity with innovation". Early Greek science displays the same pattern at work. Yet the science of the Greeks was both very much like the modern science and very much unlike it at the same time. If we take science as a certain intellectual activity, a mode of research and a historical sequence of ideas, there is a direct continuity and no essential difference between the science of Thales and his followers, on the one hand, and the science of the modern age, on the other. Copernicus and Kepler addressed the same problem of retrograde movement of the planets as Eudoxus, Hipparchus, and Ptolemy; a related problem of inequality of seasons was recognized already by Thales. As modern science, Greek astronomy knew its "normal" phases and its "crises."[11] But if we approach science from a sociologist's point of view, the difference is enormous. Greek science (not only the early) was never institutionalized, nor was it a part of an educational system. In respect to his social situation, a prominent Greek scientist can be properly described, in the words by Ben-David (though used by him for another cultural milieu), as "a charismatic personality who was honored and supported as a unique phenomenon and did not earn his living as a member of an established intellectual occupation" (Ben-David 1991, 304). Should we say, then, that Greek society just tolerated, and sometimes enjoyed, science and scientists? Should we conclude that science acquired no social function in Greek world? I don't think so.

Let me repeat that Thales and most of his followers were doing both science and philosophy. They had performed a major intellectual revolution. They had replaced the world governed by unpredictable supernatural agents by the world governed by immanent laws of nature. It was a quiet revolution.

Save for Xenophanes, the Presocratics did not attack traditional religious beliefs. Some of them, not unlike the scientists of the late seventeenth century, seem to have been pious people in a way. Some of them, like Pythagoras, even managed to base on new ideas the aspiration for immortality of the souls. Yet their research and their arguments made it evident (more or less) that the sun and the moon were just physical bodies and that all events in nature could be explained as natural events. Now, the Greek religion had been formed in a world significantly different from that in which the Greeks lived in the sixth and the following centuries. Their public life, their economy were organized in quite a rational and sophisticated way. For many of them, especially those who belonged to the upper stratum, science offered a more plausible, more adequate *Weltanschauung* than traditional beliefs did. Open conflicts between public opinion and science (or philosophy) were few; they took place mostly in democratic Athens and in particular circumstances. In general, growing science and religious practices peacefully coexisted. Scientific ideas and scientific research are referred to in the works by Euripides, Thucydides, and Aristophanes (whose attacks on science and scientists only testify to the interest in them among the educated Athenians). Science did not enter the schools, yet the sophists disseminated scientific knowledge in fifth-century Athens and elsewhere. Science provided an essential supplement, if not a really substitute, to the traditional outlook. That was its social function.

Notes

1. The only ancient civilization other than Greek, from which we have a record of competing cosmological theories, is China. The most comprehensive account of such theories seems to remain an unpublished doctoral dissertation by Christopher Cullen (1977); see also Cullen (1996). However, there are many signs of Greek influence upon ancient Chinese cosmology, which makes problematic its independent formation; see: Panchenko (2000a, 2002b).
2. The statements of this paragraph are based on Panchenko (2005, 59–77, 114–21). For a comprehensive collection of testimonies about Thales, see: Wöhrle (2009).
3. For Anaximander, in general, see: Kahn (1964/1960), Couprie et al. (2002); for the position of the earth: Bodnár (1992), Panchenko (1994).
4. Some ancient authorities attribute this great idea to Pythagoras, and some modern scholars maintain that it was in fact formulated much later, in the fourth century BC. But see: Burkert (1972/1962), Panchenko (1997).
5. On the significance of Thales' geometry, see: Mittelstraß (1974), Mittelstraß (1970, 18–32).
6. This was repeatedly denied, but see: Zhmud (1997, 157f, 163f).
7. See the first section of my paper "Thales and the Origins of Theoretical Reasoning" (Panchenko 1993), or the first chapter of my Greek book (Panchenko 2005).
8. The best general introduction to the evidence for the views of early Greek philosophers and scientists, the so-called Presocratics, is still Guthrie (1962, 1965); see also Kirk et al. (1983). For early Greek Astronomy, see: Heath (2007/1913); Dicks (1970), though valuable in several respects, is misleading for the period under consideration. For early Greek geometry, see: Heath (1981/1921).
9. On Vernant's approach, "the birth of natural thought" appears to be essentially a change in images, following a change in the environment, and not the formation of the new mode of reasoning about nature. For a more flexible and multiple treatment of the political dimension in the development of Greek science and philosophy, see: Lloyd (1979, 1989).
10. The original Russian edition was published in 1985.
11. See: Panchenko (2000b) where the Kuhnean notion of crisis is somewhat modified.

References

Bodnár, I. M. "Anaximander on the Stability of the Earth." *Phronesis: A Journal for Ancient Philosophy* 37 (1992): 336–42.

Burkert, W. 1972/1962. *Lore and Science in Ancient Pythagoreanism (Weisheit und Wissenschaft: Studien zu Pythagoras, Philoloas und Platon).* Cambridge, MA: Harvard University Press (Carl).

Couprie, D. L., R. Hahn, and G. Naddaf. 2002. *Anaximander in Context: New Studies in the Origins of Greek Philosophy, SUNY Series in Ancient Greek Philosophy*. New York: State University of New York Press.
Cullen, C. 1977. *Cosmographical Discussions in China from Early Times up to the T'Ang Dynasty*, Ph.D. thesis, School of Oriental and African Studies, University of London.
______. 1996. *Astronomy and Mathematics in Ancient China: The Zhou Bi Suan Jing*. Cambridge: Cambridge University Press.
Dicks, D. R. 1970. *Early Greek Astronomy to Aristotle*. London: Thames and Hudson.
Guthrie, W. 1962. *A History of Greek Philosophy: The Earlier Presocratics and Pythagoreans*, vol. 1. Cambridge: Cambridge University Press.
______. 1965. *A History of Greek Philosophy: The Presocratic Tradition from Parmenides to Democritus*, vol. 2. Cambridge: Cambridge University Press.
Heath, T. L. 1981/1921. *A History of Greek Mathematics: From Thales to Euclid*. New York: Dover Publications.
______. 2007/1913. *Aristarchus of Samos: The Ancient Copernicus*. Oxford: Adamant Media Corporation.
Kahn, C. H. 1964/1960. *Anaximander and the Origins of Greek Cosmology*. New York: Columbia University Press.
Kirk, G., J. Raven, and M. Schofield. 1983. *The Presocratic Philosophers: A Critical History with a Selection of Texts*. Cambridge: Cambridge University Press.
Lloyd, G. 1979. *Magic, Reason and Experience: Studies in the Origin and Development of Greek Science*. Cambridge: Cambridge University Press.
______. 1989. *The Revolutions of Wisdom: Studies in the Claims and Practice of Ancient Greek Science*. Berkeley, CA: University of California Press.
Meier, J., and A. Petersen, ed. 1998. *Karl L. Popper, The World of Parmenides: Essays on the Presocratic Enlightenment*. London: Routledge.
Mittelstraß, J. 1970. *Neuzeit und Aufklärung: Studien zur Entstehung der neuzeitlichen Wissenschaft und Philosophie*. de Gruyter.
______. 1974. *Die Möglichkeit von Wissenschaft*. Suhrkamp.
Panchenko, D. "Thales and the Origins of Theoretical Reasoning." *Configurations* 1 (1993): 387–98.
______. "Homoios and Homoiotes in Anaximander and Thales." *Hyperboreus* 1 (1994): 28–57.
______. "Anaxagoras' Argument against the Sphericity of the Earth." *Hyperboreus* 3 (1997): 175–78.
______. 2000a. "Greek Origin of the Notion of the Celestial Sphere in Chinese Cosmology." In *Syssitia: In memoriam Ju. V. Andrejev*, 174–184. St. Petersburg.
______. "On Copernicus' Success and Aristarchus' Failure." *Antike Natur wissenschaft und ihre Rezeption* X (2000b): 97–105.
______. 2002a "Eudemus Fr. 145 Wehrli and the Ancient Theories of Lunar Light." In *Eudemus of Rhodes*, vol. 11 of *Rutgers University Studies in Classical Humanities*, edited by I. Bodnar and W. Fortenbaugh, 323–36. New Brunswick, NJ: Transaction Publishers.
______. "The City of the Branchidae and the Question of Greek Contribution to the Intellectual History of India and China." *Hyperboreus* 8 (2002b): 244–55.
______. 2005. *Thales and the Origins of Theoretical Reasoning [in Greek]*. Athens.

Wöhrle, G., ed. 2009. *Die Milesier: Thales*, vol. 1 of *Traditio Praesocratica*. Berlin: de Gruyter.
Zaicev, A. 1993/1985. *Das griechische Wunder: Die Entstehung der griechischen Zivilisation*. Konstanz: Uvk.
Zhmud, L. 1997. *Wissenschaft, Philosophie und Religion im frühen Pythagoreismus*. Berlin: Akademie-Verlag.

[illegible]

4

Competition and Quality Orientation in the American, English, and German University Systems

Claudius Gellert

Starting from the thesis of Joseph Ben-David that the paradigmatic character of the German universities in the nineteenth century was largely linked to the expansion of—and the implicit competition within—that system of higher education, this chapter will lay out (1) the historical development of the American, English, and German universities and (2) account for the present-day situation of those models. In the historical overview, the functional differences in the university traditions of England and Germany shall be analyzed and, consequently, the internal differentiation of the American research universities will be explained in relation to those traditions.

Against this background, the above-mentioned thesis of Ben-David shall be evaluated, in particular regarding the role of competition rooted in—or ascribed to—the decentralized German university system. With this in mind the present deficiencies of the German universities—frequently characterized by a lack of competitive mechanisms—are analyzed in comparison to the situation in England and the United States and in relation to their historical developments. One of the aspects looked at is the recent attempt by the German government to create "elite" universities through public steering within the framework of a so-called "excellence initiative," comparable to the American and English examples.

Historical Perspectives

While universities, since the Middle Ages, have primarily been places of advanced professional training (mostly for the administrative and legal system, the churches, and the medical sector), the teaching contents during the early centuries consisted of canonized knowledge which was constantly re-interpreted, but hardly changed through empirical input. The collection and systematic analysis of empirical and scientific findings developed only gradually during the eighteenth century, and in most cases, like in England, by interested individuals outside universities. The universities themselves had degenerated into rigid, denominationally controlled organizations with limited intellectual and innovative capacities. It was only towards the end of the eighteenth century that, as a consequence of the process of Enlightenment and related political debates and changes, the universities began to become centers of intellectual discourse and critical questioning.

Beginning with Francis Bacon's critique of the production of "idola" (ideologies) by "altar and throne" (church and monarch), there emerged a rapidly growing interest in collecting data and information on all aspects of the natural and social worlds. This was a radically different approach to understanding reality from anything before. The radical paradigmatic break with the past, which happened towards the end of the eighteenth century and the beginning of the nineteenth century in several European systems, consisted precisely in the abolition of the principle of gaining knowledge through philosophical and purely abstract re-definitions and re-interpretations of concepts and ideas which was the characteristic of philosophical thinking at the time of Plato and of Scholasticism during the Middle Ages (Gellert 1992). However, even at these early stages of the emergence of empirically oriented investigations, the scope and methodology of analysis was still rather limited and often circumscribed by traditional patterns of scholarship.

In the German-speaking realm, the situation changed dramatically under the influence of the philosophers of Idealism and the Prussian administrators who were responsible for the fundamental reform of the universities in the beginning of the nineteenth century. For them, the training of students to become civil servants, teachers, doctors, etc., had to take the form of a seemingly purpose-free process of searching for truth. This required, on the one hand, a large degree of independence for the universities from state interference

(Winkler and Brachmann 2000); and, on the other hand, it presupposed an internal reorganization of universities in such a way that students and professors could pursue an understanding of "objective truths" in a combined effort. Wilhelm von Humboldt, who called this aim *Bildung durch Wissenschaft* (education through science), was convinced that the traditional relationship of authority between pupils and teachers had to be replaced by the undirected and free cooperation between students of different levels of knowledge: "Therefore, the university teacher is not any longer teacher, the student not any more just learning, but the latter researches himself and the professor only directs and supports his research" (von Humboldt 1920, 261).[1] In contrast, for instance, to England, not the student, but the subject was to receive primary attention. As Humboldt put it, "The relationship between teacher and student [. . .] is changing. The former does not exist for the sake of the latter. They are both at the university for the sake of science and scholarship" (von Humboldt 1964a, 256). Since the search for truth was not to be restricted by considerations of time, immediate occupational purposes or state control, professors as well as students had to be enabled to teach and learn what they were interested in. While this led to the students' "freedom of learning" (the freedom to select freely from what was offered in various disciplines, to change universities whenever they liked, and to take their final exams when they felt ready for them), it also had the major consequence for university teachers that an interest in new—that is, the discovery of "objective"—knowledge among students (the "co-researchers") became an important aspect of their professional (self-) definition.

One of the most essential conceptions which followed from these considerations, was the idea of a functional unity of teaching and research, that is, the teaching contents were supposed to be a direct result of the professors' research. The concept of the "unity of research and teaching" may thus be described as the normative expectation that the professional role of academics at universities should be defined in such a way that the occupational aspect of teaching is closely intertwined with, and directly based on, the ongoing process of research of the individual academic. The idea, in its original form, not only maintains that university teachers should be involved in research at all, but that the specific insights and outcomes of their respective research activities should directly become the substance and contents of their teaching. Within this tradition of thinking, the question whether or not university teachers had to be researchers was not only to be confirmed in general terms (in the sense that any research would do), but with a strict demand for immediate utilization of research results for teaching purposes (Gellert 1993a).

The actual development of the German universities during the nineteenth century confirmed in some respects the intentions of the early reformers and proved their concepts to be successful. As a consequence of the rapid industrialization and the emerging imperialism of the German Empire, the universities became true research universities. There was a permanent demand for results in fundamental research, particularly in the natural sciences, not least for military purposes.[2] The powerful university professors (Ordinarien) were engaged in a continuous process of redefining the frontiers of knowledge. The fields of knowledge were constantly changing and expanding. There seemed to be no need for a clear-cut definition of established scholarly results or for specific university curricula. In this situation, the principle of a unity of research and teaching was a natural consequence, since the training of the scholar-students followed the permanent flow of results in fundamental research.

The success of the German universities in specialized scientific and scholarly research even led to a significant influence on other systems of higher education. As Joseph Ben-David has observed, "until about the 1870's, the German universities were virtually the only institutions in the world in which a student could obtain training in how to do scientific or scholarly research" (Ben-David 1977, 22). At the end of the nineteenth and during the first decades of the twentieth century, many American and British scholars traveled to Germany. If not the function of research as such, at least the notion that university teachers should be actively engaged in research, was introduced to American and English universities to some extent under the influence of the German example.[3] Joseph Ben-David and Awraham Zloczower have, in their famous article on "Universities and Academic Systems in Modern Societies," attributed this success of the German universities during the nineteenth century to the fact that the German-speaking realm was divided up in many states, many of which had their own universities already before the reorganization of the University of Berlin in 1809–10 by Wilhelm von Humboldt, or founded new ones during the following decades. This lack of centralization of the German university system, contrary to what existed, for instance, in France and England, brought about a fundamental competition among the German universities that made them more responsive to outside requirements and also made them more willing to introduce new subjects, research areas, or innovations:

> The university system was decentralized and competitive. Universities tried to outdo each other or [. . .] had to keep pace with each other. As a result innovations were introduced in Germany more easily and accepted more widely than elsewhere (Ben-David and Zloczower 1991/1962, 130).

This upsurge, "especially in the natural sciences," they claim, "occurred not as a result but in spite of these [Humboldtian] ideas [. . .] This quite unintended success was the result of the decentralization of the German academic life. The universities, competing with each other, had to follow the successful example established in one university" (Ibid., 132). "[. . .] It was the decentralized nature of the system and the competition among the individual units which brought about the rapid diffusion of innovations and not the internal structure of each unit, or the dominant philosophy of education" (Ibid., 135).

Modifications of the English, American, and German Models

The ways in which the idea of a "unity of research and teaching" was introduced to English and American universities is a major indication of the fact that the German ideal has by no means become a universal principle. In the English case, there prevailed a strong tradition of orienting university education to the personal development of the student rather than to disciplinary requirements, as in Germany. Although the old ideal of character formation was transformed into the concept of "liberal education," which put considerable emphasis on scientific and academic training, the intellectual aspect of learning always remained embedded in the broader function of improving an individual's personality (Gellert 1988, 27f). Because of this, the English universities were able to do both: to define clear areas of established knowledge, which were organized as binding curricula, and to encourage their academics to engage in research as part of their defined duties. The latter did or did not coincide with the teachers' topical teaching program. There was no obligatory link between the two. Research became an important professional characteristic of university teachers; but the research results did by no means have to be directly utilized for teaching purposes. In this way, it was ensured that academics became and remained acquainted with research activities. This seemed to be a sufficient proviso to guarantee high intellectual standards in teaching.[4]

In the United States, the German research example was also adopted towards the end of the nineteenth century. But there, in contrast to England, the consequences for the organization of university teaching and research were more radical. On the one hand, there occurred a complex process of differentiation in the overall system of higher education, which was related to diverging interests, purposes, and functions in tertiary education (Riesman 1980; Veysey 1965). The sector which is comparable to European universities, that is, the "research universities" and the doctoral institutions, on the other hand, was characterized by a gradual process of organizational and functional segregation from within.[5] The three major functions of the leading American universities today correspond to a threefold structural segmentation: First, the function of liberal education, in many respects similar to its British counterpart, is almost exclusively reserved for the undergraduate level.[6] Second, the function of professional training is placed in specialized professional graduate schools. And finally, the research function is exercised mainly within the academic graduate schools of arts and science. The American research universities are, thus, characterized by an almost complete segmentation of teaching and research, at least if compared to the German university tradition. Of course, the professors there are often the same, whether they teach undergraduates or graduates, whether they deal with PhD-candidates or with aspiring professionals. But, a close connection between ongoing research and teaching exists primarily in the graduate schools of arts and science[7]; this is the sector within the American research universities which has preserved—and developed—the German heritage (Gellert 1993b).

In Germany, in contrast, the expansion of the university system after World War II has led to an awkward structural and functional muddle. The transformation of the system into places of mass higher education (Trow 1970), with about six or seven times as many students now as in the early 1960s, has jeopardized the traditional balance between the tasks of academic inquiry and advanced training of students. The old ideal of a unity of research and teaching is still part of the official value frame of reference at universities. But in recent decades there occurred frictions in this system because of an increasing discrepancy between the traditional research orientation of university teachers and their factual involvement in professional—or even vocational—training of large numbers of students.

This tension not only occurred in Germany, but all over Europe, and has been threatening to break up the traditional institutional framework, either by external (overall) differentiation (e.g., the introduction of *Fachhochschulen* in Germany, the IUT's in France, or the polytechnics in Britain, that is, alternative forms of tertiary institution which were more vocationally and practically oriented than universities), or by internal functional diversification (e.g., attempts to distinguish professional from research programs) (Gellert 1991). But despite several decades of reform discussions, the German model has until very recently—when, due to the Bologna process, BA and MA-programs were introduced on a massive scale—been characterized by antagonistic structural features: on the one hand, the students' ability to freely choose subjects, universities and their time of examination; on the other, the professors' freedom to teach whatever they like (both sanctioned by the Humboldtian principle of the freedom of teaching and learning, referred to above); the constitutionally guaranteed open access to all universities for anybody with a respective secondary degree; the bureaucratic and state control of all curricular and organizational matters, including the civil service status of the professoriate; the over-loading of programs and courses according to individual research interests of the professors; and finally, the widely criticized actual length of studies (i.e., time to degree) in most subject areas.

This process of transformation into a system of transparent organizational structures still continues, and whether the final result will be much better than what existed before, is by no means clear (Gellert 1999). What is clear, however, is the fact that one of the most fundamental deficiencies of the German higher education system in recent decades was the lack of independence from state control—and thus, among its actors, a lack of comparing their aims, activities, and achievements also with other universities and their achievements. The division of the country into many states which were, and are, almost totally autonomous in all educational and university matters, has clearly not sustained what Ben-David and Zloczower identified as a major result of decentralization in the German realm during the nineteenth century, namely, competition within the institution of higher education. Very slowly now, there is insight growing in the public sphere that an awareness of the activities at other universities and departments, and a constant comparison with those, is a crucial prerequisite for quality-pursuit on all levels of the organizations. Decentralization as such certainly has not done the trick.

Institutional Differentiation, Competition, and Quality Orientation

Under the influence mostly of the US and UK paradigms, there is a growing awareness in continental Europe that functional differentiation, that is, the pursuit of differing tasks and roles in higher education on a high quality level, can only be maintained if a dynamic structural hierarchy of quality and prestige distinctions between sectors, institutions, and departments is accepted as a normal phenomenon. Increasingly, the examples of higher education systems in the United States or Britain are getting academics and policy-makers in Europe to realize that exactly measuring quality differences is not necessary for achieving institutional quality orientation. The main point is that in those countries, despite the development of "sophisticated" methods of quality assessment (like the RAE in the United Kingdom), the overall process of classifications of institutions and departments is based on individual peer-group evaluations, and the differences between sectors and universities which are perceived through presentations in the media, have their own dynamic, insofar as they result in an overall process of control and incentive mechanisms on all levels.

It is the awareness of differences among institutions, above all, which makes individual professors, lecturers, department heads, deans, and university presidents constantly strive for improved quality—from the selection of students to appointments and to promotions. The question whether these differences are really exactly measured is of secondary importance. What matters is that everybody involved in the system accepts, as quasi natural, that quality and prestige differences exist within the higher education system—or within a particular sector such as the research universities. Whether such results can in a sustained perspective be "artificially" brought about by government intervention and selective financial support is, however, at least questionable.

In recent years, it has also dawned upon German politicians that competition and quality orientation among universities is of vital importance if they are to be internationally recognized players in the scientific community. Slowly, a discussion is developing which addresses, for instance, the question to what extent it might be necessary to give universities a larger degree of autonomy in order to enable them to act more freely in matters concerning research or teaching. However, for politicians who are in need of short-term results—because elections are lurking around the corner—it is difficult to accept that processes, which have brought about Oxford, Cambridge, Harvard, or Princeton, take centuries.[8]

Thus, the Federal government came up with the idea to persuade the education ministers of the German states (*Länder*) to agree to a program which is supposed to create the equivalent of Yale or Berkeley within a few years. Only a few years ago just under two billion Euro were offered for German universities to be spent over a period of five years, provided they were willing to bid for some of the money with conceptions for high-quality graduate schools, research projects in collaboration with industries, or generally ideas on how to become an internationally recognized elite university.[9]

The reactions in the public have been very divergent. While politicians and some of the universities reacted rather positively, other comments were critical. First of all, for many observers it was by no means clear why the few universities, who had proven for a long time that they were able and willing to organize high-quality research, now suddenly needed extra money to do exactly that. But even more fundamental was the critique that the majority of the universities in Germany, which were not elected to this elite circle, would have to live during the coming years with the stigma of belonging to a "secondary class" of institutions which, in turn, would likely hamper also their future attempts to gain outside research funding, and that, in the end, could make it nearly impossible for those institutions to be successful, if there should be another round of this Excellence Initiative (Gellert 2008; Pichlbauer and Rosner 2008). A different kind of fundamental critique suggested that, in reality, we are dealing in Germany with a kind of monopolistic and oligarchic structure of research funding, in which the leading funding organizations like the German Research Foundation (DFG), which are dominated by a small number of (mostly southern German) universities, determine what counts as valuable research; this is not only detrimental for the majority of institutions which are left out when it comes to the distribution of funds, but is also counter-productive for the innovative potential of research in Germany in general (Münch 2007).

The most decisive aspect from my perspective, however, is the fact—pointed out before—that we are dealing in America, and also in England, with institutional processes which comprise all research universities, and which have led on all levels to competitive mechanisms and to a permanent achievement orientation. It is only as a consequence of this all-comprising and continual process, in which all respective organizational units are participating, that *inter alia* there emerge elite universities simply because in an evaluative hierarchy, some are at the bottom, some at the top, and some are in between. As long as everyone has a chance to improve the relative position in the hierarchy, the system as a whole functions beneficially for all. The Excellence Initiative is exactly the opposite of the situation in the United States. It mostly reflects the unbroken determination of those in charge in the Federal and State governments not to loosen their grip on the institution of higher learning; and marks also a kind of preservation of continuity and tradition in Germany (Gellert 1993b).

Conclusion

The debates on university reform, which have been going on in continental Europe for more than forty years, have often been influenced by references to the American and British models. In terms of quality control with regard to research and teaching, the practical consequences, so far, are still rudimentary. The Bologna process, with the introduction of BA/MA degree programs, might be a major step in the right direction. Increasingly, the awareness among academics and policy-makers is growing that functional differences in the overall higher education system have, by necessity, to be accompanied by quality and prestige differentials. Internal evaluation mechanisms (not the least with regard to teaching) in European universities, which were introduced in recent years, are an important indication of this. More essential, however, are developments in several European countries (e.g., in Austria) towards more autonomy of organizational units within their respective higher education systems.

Germany, in this respect, is lagging behind. The basic idea that competition is a vital prerequisite for the enhancement of quality in research and teaching, as was suggested in several of Ben-David's contributions, is still of fundamental importance. Ben-David and Zloczower argued that rivalry and competition in an expanding system was directly linked to the success of the German university during the nineteenth century, but once a certain degree of saturation was reached and expansion slowed or came to a halt (early in the twentieth century), competition alone could not be properly practiced or had lost its force: due to its structure and internal organization, the German university—in contrast to the US university—was unable to sustain a growth path (Ben-David 1991, 129). Furthermore, it may well be argued that the "success" of Germany's universities at that time, that is, their international reputation as efficient research institutes, came already bundled with a tremendous counterproductive side-effect, namely, the political and cultural contribution as seats of authoritarian, racist, and undemocratic ideologies (Gellert forthcoming). Insofar, the present attempt by the German government to create "elite" universities, comparable to those of Cambridge or Harvard, will reinforce the interest and power structures of the German research system; but it will not bring about overall competitive mechanisms within academia nor an improvement of creativity and innovation in the universities.

Last but not least, it should be emphasized that elite-oriented, competitive, and open academic systems like those in the United States or England, are also characterized and guided by strong normative and ethical principles. One of the main aspects here is that the "knowledgeable agents" (Giddens 1984) in those systems are usually not prepared to separate between areas of relevance with regard to ethical considerations. Everyone is normally accepting to be measured in his or her actions against the value system which prevails. While the political correctness, particularly within the United States, may sometimes appear to be slightly exaggerated, it is probably correct to say that more of that ethical orientation may also be appropriate in some German contexts. For instance, in conversations with academics of some leading English and American universities it became crystal clear that it would be considered unacceptable that a university president, who has been officially convicted of massive tax fraud, could continue in this function at their universities—or that he could even be re-elected by his fellow professors. But that is what happened in one of the new German elite universities. The only question remaining is, who is to blame in a situation like this? The president, hanging onto his job? His professorial colleagues for supporting him? Or the politicians and research organizations who have, nevertheless, elected that university for being (together with two other institutions) the "best" in Germany?

Notes

1. Translation from German, unless otherwise indicated, is by Claudius Gellert.
2. It has been calculated that roughly one-third of all financial support of university research at the end of the nineteenth century was spent for military purposes; cf. Prahl (1978, 227f).
3. This applies less to Scottish universities which had a strong research tradition. For the United States, cf. Diehl (1978); for the English case, cf. Ashby (1967).
4. One indication of the growing importance of research as a necessary qualification of academics was the introduction of the PhD in the early twentieth century (Simpson 1983).
5. Nowadays, there are a bit more than 200 doctorate-granting institutions in the United States, including "research universities" (see http://classifications.carnegiefoundation.org/resources/).
6. The American "liberal arts" concept is, however, characterized by a stronger interdisciplinary emphasis, while in England the "Single Honours Degree," that is, a specialized training in one subject, is still prevalent.
7. The connection can also be observed in fields of engineering, for instance.
8. Stanford University might be an exception, but even there it took close to a century to transform a provincial institution into a world-class university.
9. This program runs under the name of *Exzellenzinitiative;* in the meantime nine universities have been elected into the status of an "elite-university."

References

Ashby, E. "The Future of the Nineteenth Century Idea of a University." *Minerva* VI (1967): 3–17.

Ben-David, J. 1977. *Centers of Learning: Britain, France, Germany, United States.* McGraw-Hill/Transaction Publishers.

______. 1991. "Social Factors in the Origin of a New Science: The Case of Psychology." *American Sociological Review* 31 (1966): 451–65.

Ben-David, J., and A. Zloczower. "Universities and Academic Systems in Modern Societies." *European Journal of Sociology* 3 (June 1962): 45–84.

Diehl, C. 1978. *Americans and German Scholarship, 1770–1870,* volume 115 of *Yale Historical Publications.* New Haven, CT: Yale University Press.

Gellert, C. 1988. *Vergleich des Studiums an englischen und deutschen Universitäten,* 2nd edn. Frankfurt: Peter Lang.

______. 1991. *Alternatives to Universities.* Paris: OECD Publications.

______. 1992. "Faculty Research." In *Analytical Perspectives,* volume 3 of *The Encyclopedia of Higher Education,* edited by B. R. Clark, and G. Neave, 1634–41. Oxford: Pergamon Press.

______. "The German Model of Research and Advanced Education." in Clark [1993], chap. 1, 5–44.

______. 1993. *Wettbewerb und Leistungsorientierung im amerikanischen Universitätssystem.* Frankfurt: Peter Lang.

______. 1999. *The Changing Conditions of Teaching and Learning in European Higher Education.* In *Innovation and Adaptation in Higher Education* (ed. By Gellert) *Higher Education Policy Series,* 22, 9–30. London: Jessica Kingsley Publishers.

______. 2008. "Ethik, Eliten und die Rolle der Universitäten." In *Systemdynamik und Systemethik. Verantwortung für Soziale Systeme: Gedenkschrift für Ludwig Bühl,"* Pichlbauer, M. and S. Rosner, 223–233. Munich: Rainer Hampp Verlag.

______. Forthcoming. *Society, Politics and Universities in England and Germany.* Frankfurt: Peter Lang.

Giddens, A. 1984. *The Constitution of Society: Outline of the Theory of Structuration.* Berkeley, CA: University of California Press.

Munch, R. 2007. *Die akademische Elite: Zur sozialen Konstruktion wissenschaftlicher Exzellenz.* Frankfurt: Suhrkamp.

5

Organizational Behavior and Strategic Change of Public Research Institutions in Turbulent Scenarios

Mario Coccia

Introduction

Current scenarios based on market volatility, a changed political climate, economic crises, budget retrenchments, a growing interest for technology transfer, and so on, have led governments in most countries to reorganize public research bodies (PRBs) and universities along the lines of a new public management reform (cf. Coccia 2009a, 2009b; Coccia and Rolfo 2010, 2007). In fact, the core of the debate pertains to a national system of innovation,[1] made up of solid research labs and universities (Herbst 2004, 2007), that is capable of generating scientific findings and technical know-how and supports innovation and education. The objective is to increase efficiency and effectiveness of public research structures, in order to foster the transfer of technological services and applied research findings to public institutions and private companies—as well as to the overall economic system (Dresner 2002).

The aim of this chapter is to analyze the "strategic changes" (Gioia and Chittipeddi 1991) and the resulting new morphology (Ben-David 1991; Herbst 2004, 2007) of the largest Italian public research body that was brought forth by public management reforms. The lesson learned can be useful to understand the modernization of PRBs and their associated organizational behavior to cope with a turbulent economic environment.

Before discussing these issues, let me (1) describe the main European and US scientific research bodies that provide a basis to analyze the evolutionary changes of public research institutions and (2) investigate strengths and weaknesses—as well as threats and opportunities—of their new morphology within the national system of innovation.

Public Research Bodies in the United States and Europe

The United States are often considered a leading nation in research and innovation, not the least because of the setup of their national system of innovation that includes modern, competitive research laboratories (Ben-David 1972). The universe of American research labs is complex and includes some 700 federal laboratories directly financed by the US government, plus hundreds of university research labs as well as almost 14,000 industrial labs (Crow and Bozeman 1998). There are also laboratories that exhibit hybrid characteristics and cannot be strictly included in any of the above three categories. University labs are often part of larger institutions with a broader spectrum of missions, some of which are not strictly either academic or research oriented. Moreover, the American national system of innovation has a well-established tradition of large research laboratories, such as the Bell Labs, RCA's David Sarnoff Laboratory in Princeton, Fermi Lab, Lawrence Berkeley National Labs, and the Los Alamos National Laboratory. For instance, the Los Alamos National Laboratory, established in 1943, has 7,000 research personnel from the University of California and over 3,500 researchers hired from the Department of Energy. Los Alamos is subdivided into ten directorates covering a total of some fifty divisions, programs, or offices.

All US laboratories are grouped into departments, divisions or branches, and finally into research teams, led by a principal investigator,[2] more or less *ad hoc*, based on the needs of the projects under investigation. Many laboratories are organized on more than one basis. University research labs are most often organized according to a principal investigator model, whereas government labs are more commonly organized on the basis of departments and principal investigators (matrix organization); these are often considered outstanding models of scientific efficiency (Crow and Bozeman 1998).

Europe, in comparison, has several public research organizations, the most important of which are the *Max-Planck-Gesellschaft* (MPG) in Germany, the *Centre National de la Recherche Scientifique* (CNRS) in France, the *Consejo Superior de Investigaciones Científicas* (CSIC) in Spain, and the *Consiglio Nazionale delle Ricerche* (CNR) in Italy. These bodies are further divided into a large number of institutes and laboratories supported by a central administration and by scientific coordination structures (for instance, departments in the French case and a senate in the German model).

The structure of the *Max-Planck-Gesellschaft* in Germany is made up of a central organization and a network of associated research institutes. The organization deals with aspects of governance as well as public relations and liaises with the government in the bargaining for public funds assigned annually to the research organization. Its scientific network is composed of about 79 institutes with a total of 12,000 personnel (3,200 of which are researchers), plus 9,100 doctoral and post-doctoral students as well as associate researchers or assistants. Its scientific activities are grouped into three macro-areas: biology and medicine; chemistry, physics, and technology; and humanities and social sciences. Similar to other major German research bodies (*Fraunhofer-Gesellschaft, Leibniz Gemeinschaft, Helmholz-Gemeinschaft*), the *Max-Planck-Gesellschaft* is characterized by comparatively large individual institutes, each of which is subdivided into smaller research units focusing on specific tracks or lines of research. The resources of the MPG are almost entirely composed of public funds (around 94 percent).[3] Activities carried out on behalf of non-governmental clients generate a relatively small part of proceeds: the relationships with the industrial system as well as technology-transfer activities are dealt with by an external affiliate company, the *Garching Innovation GmbH*, and by the *Fraunhofer-Gesellschaft*. Analyzing the German MPG, Leilich (2005) shows that the institutes have two main types of a structural setup: flat hierarchies, in which all researchers work directly with the director of the institute; and steep hierarchies where senior researchers form intermediate layers of control between the director and the junior researchers. Institutes specializing in the natural sciences, characterized by a high level of multidisciplinarity, display a steep hierarchic structure; conversely, institutes specializing in humanities and social sciences, with little multidisciplinarity, tend to have a flat hierarchical organization.

In comparison to the German situation, the French model is highly fragmented: the *Centre National de la Recherche Scientifique* (CNRS) is made up of 1,257 laboratories with around 26,000 personnel, that is, permanent staff (11,000 of which are researchers), divided into six departments: mathematics, physics, earth sciences, and astronomy;

chemistry; life sciences; humanities and social sciences; environmental science and sustainable development; engineering; and two National Institutes: the National Institute for Nuclear and Particle Physics, and the National Institute for Earth Sciences and Astronomy. CNRS's body is based on the organizational concept of *unité mixte de recherche* (Larédo and Mustar 2004). These so-called mixed units are research groups of small to medium size and located in university buildings, or on campuses with which they have strong interactions (in relation to the presence of university professors and researchers); the CNRS-laboratories involve also a total of 60,000 students, professors, and external researchers (apart from the permanent staff). Eighteen regional delegations deal with the management of all the laboratories and take care of fostering relationships with local subjects (e.g., local authorities, universities, firms, other institutions, etc.).

Spanish *Consejo Superior de Investigaciones Científicas* (CSIC) has a structure that lies in between the German and the French model. It is made up of ninety-four institutes, twenty-nine centers and twenty organizational units. There are 149 research groups with over 6,000 personnel (2,300 of which are researchers and 4,000 of which are trainees) divided into eight different scientific areas (cf. Coccia and Rolfo 2007, 2008, 2009, 2010). This overview is the background to describe, compare and analyze the Italian case study.

New Morphology of Italian Public Research Institutions

The Italian National Research Council (CNR) is similar to other European research institutions; it is a public body whose task it is to carry-out, to promote, to spread, and to enhance research activities aimed to foster the scientific, technologic, economic, and social development of Italy. Its scientific activities are divided into the following research fields: biomedical sciences, physics and technology of the matter, science of earth system and technologies for environment, chemistry and material technologies, engineering, information and communication technologies, technologies for energy and transportation, biological and food sciences, socio-economic sciences, human sciences and cultural heritage. The CNR operates within Italy through a wide network of institutes and decentralized units; this facilitates a diffusion of scientific competences and fosters collaborations with local institutions and businesses. The operation of CNR is based on a three-year activity plan (updated yearly) that sets up general guidelines, objectives, priorities, and resources for the short and long run, consistent with the National Research Program and with the research programs of the European Union. The three-year plan specifies the financial requirements, that is, the budget, spells out the implicit human resources requirements and personnel recruitment needs in the respective scientific fields, schedules projects, and carries out the planned scientific activities. The Italian "Ministry of Education, University, and Research" must approve this plan and its yearly updates.

Table 5.1 compares financial and human resources as well as scientific productivity measures by major European scientific institution, using data of 2002 (the last available data in the CNR Report of 2007). The results show that the German MPG commands over the largest financial resources and produces the most publications per researcher.[4]

At the end of the 1990s, the debate concerning public research (and the CNR) in Italy focused on its modest economic impact and the low productivity of the research units (cf. Coccia 2001, 2004, 2005a, 2005b, 2008a, 2008b). The causes were to be found in a presumed outdated morphology (Ben-David 1991; Bourdieu 1988/1984) of the research institutions, with units judged to be too small to use the necessary infrastructure to capacity; associated with that there were scale diseconomies and high unit costs of capital equipment usage, maintenance, and administration. To fight such deficiencies, the Italian government launched in 1999 a public management reform with the aim to consolidate and merge about 310 research units into roughly 100 new institutes (see Figure 5.1). The notion of increasing the returns of scientific research by concentrating financial and human resources in larger laboratories formed the foundation of the Italian research policy, a policy already implemented in the United Kingdom during the 1970s (Johnston 1993, 1995). The notion of the existence of (internal) economies of scale in scientific production is based on: a critical mass (i.e., the size) of research units below which researchers cannot activate significant cooperative relations; inseparability of some inputs; administrative activities (or research infrastructures) characterized by fixed costs with respect to the volume of the activity; and the view that larger research institutes have a higher capacity to foster and manage major projects.

However, because of the rules pertaining to the civil servants status of CNR staff, it was impossible to accommodate all the personnel physically in the new institutes, with a consequent scattering in more than 200 locations: in fact, every (virtual) institute has a main institute

Table 5.1
Comparison of European Research Bodies (Number of Researchers [Res], Research Budget [in 1,000 €], and Publications per Researcher [Pub/Res]), by Year

	Year	CNR (I)	CNRS (F)	CSIC (E)	MPG (D)
Researchers	2001	3,693	11,643	2,259	3,116
	2002	3,610	11,643	2,713	3,509
Budget	2001	214,672	211,028	178,840	404,685
	2002	225,485	217,556	207,151	357,082
Pub/Res	2001	1.34	1.42	1.93	2.42
	2002	1.36	1.39	1.89	2.19

Source: CNR Report (2007).

(headquarters) and several decentralized research units. In general, the effect of this basic reform on the structure has been: "*Plus ça change, plus c'est la même chose*." Instead, as far as organizational behavior is concerned, this territorial layout of institutes turned out to be the source of managerial and organizational strain and, at least in the short run, it did not bring about any noticeable improvement of scientific collaborations among the research units: the separated units tried to maintain part of their past autonomy (cf. Coccia 2009a, 2009b).

While this reform was still under way, the government decided in 2003 to launch a new reform, focusing this time on project management, with the intention to encourage CNR to embrace and emphasize entrepreneurship and to supply technological services and applied research activities to firms and other external subjects (cf. Coccia, 2008a, 2008b; Coccia and Rolfo 2010, 2009, 2008). The effects of this reform are ongoing. This second reform, inspired by the principles of project management (Bryde 2003; Sydow et al. 2004), is based on a matrix structure in which each resource involved in developing a project is simultaneously subject to a double line of authority: the director of institute and the project manager. This type of management structure is particularly suitable when several large projects are being managed simultaneously (Coccia 2008a, 2008b; Coccia and Rolfo 2008, 2009, 2010).

The new morphology of a project-based organization of the Italian national research council pivots on departments and institutes. The departments of the CNR's new organization are seen as back-offices

Figure 5.1
Organization Change of the National Research Council of Italy (Adapted from Coccia and Rolfo 2010)

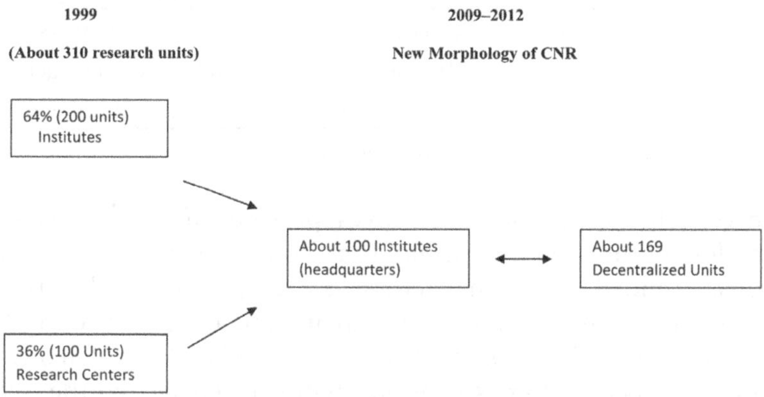

to coordinate the various institutes in the process of expanding their knowledge. Specific departments were created to cover the fields of agriculture and food, energy and transport, cultural identity, ICT, production systems, molecular design, medicine, materials, life science, earth and environment, and cultural heritage. These in turn are grouped by interdisciplinary "thematic research areas." In order to develop their projects, departments are meant to call for the contribution of several institutes and to ensure reciprocal coordination with other thematic areas; they have the task to set up the scope of the project (i.e., its major objectives and the project schedule) and provide the overall resources necessary. Departments are given the task to appoint project managers, to approve proposals put forward by those, and to supervise projects. Departments are also responsible for inter-departmental relations or relations with third parties (on a national or international level) as well as for the setup of project evaluation and peer review. In short, departments are meant to act as interfaces between CNR's supply of scientific research and expertise, and society's exigencies (in Italy or Europe in general).

To sum up, the departments have macro-objectives (projects) from which a bunch of job orders from the market branches out within the institutes. Each job order within the research lines of the institute is managed by a job order manager and can have one or several work packages, driven by work package managers, in general researchers (see Table 5.2). Job order and work package managers are the executive

Table 5.2
Organization of Project Management in CNR (year 2012)

Structure	Scientific Levels	Manager/Organizer
Department director	Projects	Project managers
Institute director	Job orders	Job order managers
	Work packages	Work package managers

Source: Adapted from Coccia and Rolfo (2009).

arms to achieve the macro-objectives (projects) of departments—and therefore of the CNR as a whole. These job orders are generally small projects funded by external actors, and the projects of departments are in many cases no more than bureaucratic boxes collecting "real" projects of a smaller scale that researchers have often been awarded competitively by a domestic or international agency. In other words, departments do not, as a rule, manage "real" projects directly: the institutes are the project-based organizations (cf. Coccia 2008b, 2009a, 2009b; Coccia and Rolfo 2008, 2009, 2010).

As far as human resources are concerned, they are subject to the authority of two lines of management: the project manager (appointed by the department director), and the institute director (see Table 5.2). Job order managers are appointed by the department director following a designation by the institute director, whereas work package managers are appointed by the institute director. On the other hand, researchers are subject to several levels of hierarchy, such as—from the administrative point of view—that of institute directors, and—from the scientific point of view—that of job order managers and work package managers. Therefore, the old organization of institutes was based on flat hierarchies, and now, with the new morphology (Ben-David 1991) focused on project management, institutes have steep hierarchies which create coordination problems between the intermediate levels of the organization and conflicts among the researchers themselves (cf. Coccia 2009a, 2009b).

However, the new morphology of the CNR does not follow properly the project management principles it espouses. Ideally, financing should be allocated competitive on a "per project" basis and no longer on the basis of a "subject" (institute). In the past, allocating funds per institute or subject matter provided a bargaining platform for institute directors and the president of the CNR. This bargaining was considered a source of inefficiency, because public funds were flowing in the direction of the most influential scientific lobbies rather than towards research units with a high performance potential or history. Although the restructuring of the Research Council was aimed at overcoming the deficiencies of the old funding scheme, today there is no competition to fund institutes because each research project or focus is tackled exclusively by one single institute or research team. The weakness of this process is the indiscriminate allocation of public resources to "dogs" and "stars" among the research institutes, generating a myopia of management based on an excessive consideration for short-term needs rather than on a long-term strategy to increase the performance and the efficiency of the overall research system (cf. Coccia 2001).[5]

Modernization and Drawbacks of Western Style Scientific Organization

The morphology of the Italian CNR has been affected by policy changes carried out in accordance with a shift in government and governance. The first reform of the CNR in 1999 was inspired by the idea to gain efficiency through the consolidation—and associated larger sizes—of research units. In particular, the purpose of the first reform was to increase the size of institutes, premised on the equation of "large labs = efficient labs," and emulating the institutes of the MPG in Germany. This change in research policy was guided by a narrow administrative—and not a scientific—stance. Although today there are about 100 newly formed, consolidated institutes (as opposed to around 310 research centers in the past), these are often composed of several decentralized units located far from the respective headquarters. Such consolidation creates diseconomies of scale due to the increased costs of coordination of decentralized units (cf. Coccia and Rolfo 2007), which in turn spawns an organizational bureaucratization with negative effects regarding research productivity (Coccia 2009a, 2009b): the relevant literature indicates that smaller rather than larger sized laboratories are more productive (cf. Carayol and Matt 2004; Coccia and Rolfo 2002, 2007, 2009, 2010).

While this reform was still under way, a new government in 2003 decided to launch a restructuring of the CNR based on project management principles, with the explicit aim to transform the CNR into an entrepreneurial body operating at the service of firms and supplying technological services for small- and medium-sized enterprises (cf. Coccia 2008; Coccia and Rolfo 2009, 2010). This second reform, which oriented itself on a matrix structure model, has caused further coordination problems due to the proliferation of channels of communication that creates informational log-jams and conflicts, not only between the respective directors of institutes and the associated departments, but also among the researchers because of the now pronounced hierarchical organization (cf. Coccia 2008b; 2009a, 2009b; Coccia and Rolfo 2007, 2008, 2009, 2010).

Because of its new orientation, the Italian National Research Council is focusing on technological services and applied activities rather than on basic research. In the past, required financial resources came mainly from the Italian government, but now the council receives quite large amounts of funds from private industry or the European Union: in fact, 50 percent of its budget is third party financed (cf. Coccia and Rolfo 2002, 2008, 2009, 2010). These funds are secured through consultancy studies and technological services for external clients, from private businesses, as well as through research grants from the European Union or other international bodies. In particular, the technological services of Italian National Research Council cover: analyses and technical tests (chemical and physical); technological services (homologation, calibration, nuclear magnetic resonance, etc.); quality services (accreditation, certification, quality control, etc.); environmental services (water monitoring, pollutant emission control, etc.); information technology services (data elaboration, supply of databases and data, etc.); health services; other research activities based on contracts with firms and institutions (cf. Coccia and Rolfo 2002).

Like the CNR, other European research institutions have opted for a project-based organization. In fact, Mangematin et al. (2006) analyze the French National Laboratory of Advanced Technologies (NLAT) which decided in 1998 to change its organization from a team-based to a project-based form (Sydow et al. 2004; Zeller 2002). Looking for explanations and motives for this change, the authors identify the following elements: low project core staffing levels which stimulate the circulation of engineers and researchers among projects and blurs project boundaries; implementing and managing thematic projects which build on specific competencies developed in dedicated projects; and encouraging "bricolage" to hybridize project management with traditional hierarchical management practices. Most NLAT engineers and technicians are not allocated to a specific project full-time, but are supposed to move from one project to another; this circulation of individuals communalizes project management practices within the organization. However, such practices tend to destroy the tacit knowledge linked to projects, and if changes are too frequent, the realization of projects can be slowed and the learning processes dissolved. While project management tools have been designed to manage specific projects,[6] the fact that they are not adapted to the management of thematic projects—as well as the fact that project management tools are not supported by top management—explains why projects are held up frequently (Mangematin et al. 2006). Project management ambiguities regarding the definition of the project and the responsibilities of project leader, etc., induce risks (Miller and Lessard 2001). Frequent delays, and inadequacy of project leaders' answers to managerial problems, appear to have produced a management crisis (Mangematin et al. 2006; Mihm et al. 2003).

In general, project management applied to public research institutions has several drawbacks. Managerial literature agrees that a matrix structure can be effectively used for *ad hoc* projects, when a larger exchange of information and a greater deal of professional relations among the project's members of staff are required and resources need to be divided up (Bartlett and Ghoshal 1990; Fassero 2004). However, developing a matrix organization in PRBs requires time as well as the willingness to take on new roles and learn new functions by public servants; all this implies major implementation difficulties in public organizations. Although the matrix organization certainly has some advantages, there are also a great deal of potential difficulties and organizational problems, as described by Mangematin et al. (2006). Mullins (1999) also stresses other problems in project-based organization, such as:

- personnel is reluctant to accept a constant change and prefers the organizational stability of their own functional grouping;
- matrix organizations cause coordination problems and conflicts among researchers;
- there are problems in defining the extent of the project/job order manager's authority over the staff from other departments and in gaining the support of other functional managers; and
- personnel tends to neglect their duties and responsibilities.

Therefore, unless the matrix structure is properly designed and implemented, it tends to become unmanageable over time and it is likely to create conflicts and confusion within public research structures. Moreover, the proliferation of communication channels leads to informational log-jams where overlapping functions cause bureaucratization as well as a loss of responsibility in relation to one's own duties (cf. Coccia 2009a, 2009b; Coccia and Rolfo 2008, 2009; Fassero 2004; Mullins 1999).

Since 2012, the Board of Directors of CNR, under the pressure of Italian governments to reduce costs and increase the efficiency, is going to reform again the structure by the reduction of scientific departments from eleven to seven units and the implementation of new organizational changes that are in progress: in essence, this new reform is also designed as "*Plus ça change, plus c'est la même chose*" since rules, mission, operation and management in all scientific and administration structure are not changed. The main short-term effect of these continuous reforms is to increase the instability of the structure and uncertainty of organizational behavior.

Strategic Change and Consequences of New Morphology of Scientific Institutions

A change involves an attempt to alter the current way of thinking and acting by the membership of an organization. More specifically, "strategic change involves an attempt to change the current modes of cognition and action to enable an organization to take advantage of important opportunities or to cope with consequential environmental threats" (Gioia and Chittipeddi 1991, 433). New morphologies of research institutions and novel tendencies of the research sector have had implications which I shall focus on in the remainder of this section.

Market-oriented research institutions: Common features of the Italian and European public management reforms are the shrinking public research budgets: public funding is not any longer sufficient to cover the current expenses of various public research units. In fact, the requirements for increased accountability, an increased flexibility of research institutions to adapt to a changing environment, and a stronger focus on socio-industrial objectives have led most countries to modify their research funding. In the past, public funding enabled Italian scientists to carry out a base-load of scientific activities and to secure additional resources, if necessary, for externally funded projects. Nowadays, because of the decline of resources provided by the State, it is impossible to conduct research solely with such funds alone. Italian research bodies consider their base funding not sufficient to maintain their structure, which means that research equipment cannot be maintained, repaired or renewed; international projects based on joint collaboration cannot be followed up; and the participation in international conferences, congresses or meetings is severely constrained. Research institutes are forced to apply for and to secure resources from industries and private foundations to conduct their normal scientific activities. "External," third party funding has become not just an additional, but the main funding source for the Italian public research institutes. Through such reforms, the government intends to foster multidisciplinary, market-oriented research, and an enhanced cooperation with the private sector. Since the year 2000, in fact, third party funding through technology transfer activities of Italian PRBs, operating in the natural sciences, has increased while public funding has been reduced (cf. Coccia 2009a, 2009b; Coccia and Rolfo 2007, 2008, 2009, 2010).

Bureaucratization within research institutions: As mentioned, the impetus for reforms regarding the Italian national research institutions is the promotion of entrepreneurial behavior through a focus on multidisciplinary, market-oriented research and the cooperation with the private sector of the economy (cf. Coccia 2008b; Coccia and Rolfo 2009, 2010). The other side of coin, however, is that the new morphology of the CNR has created problems of coordination and caused increased bureaucratization, affecting the productivity of research units (cf. Coccia 2009a, 2009b; Coccia and Rolfo 2007). This has been corroborated by 87 percent of the personnel (researchers, technicians, and administrative personnel) working in institutes of the Piedmont and Lombardy, two large industrialized regions in the North of Italy: the CNR's new structure has increased administrative duties and red tape, not the least because of unclear lines of control, the replication of administrative activities, and the management of the broad portfolio of market-oriented research projects (Coccia 2009a; cf. also Crow and Bozeman 1998).

In particular, the bureaucratization of the National Research Council laboratories in Italy is due to a hasty, poorly designed, and incomplete restructuring that generated structural deficiencies (cf. Coccia 2008b, 2009a; Coccia and Rolfo 2007, 2008). In fact, the horizontal, vertical, and geographical differentiations of large organizations need extra administrative resources to keep the institutions together. Mintzberg (1983) argues that the different activities of an organization must be coordinated and controlled, and the larger the organization, the more emphasis will be put on these tasks.

To sum up, the first restructuring of the CNR (based on consolidation and mergers) created diseconomies of scale due to higher costs of coordination of decentralized research units. The second restructuring caused administrative loads associated with the management of a broad portfolio of small research projects. The third (in progress since 2012) has creating uncertainty and instability. These heterogeneous research policies, operating side by side, encouraged administrative bloat; but it also appears that a significant part of this bloat is due to duplications along the vertical and horizontal dimension in the organization. The cost of this bureaucratization is invisible in the short run. Moreover, the second restructuring with its explicit aim to transform the CNR into an entrepreneurial body, is based on a misguided conception of a performance-oriented governance which affects public research institutions (Herbst 2007). In effect, bureaucratization tendencies are a direct cause of a badly designed matrix-structure (Coccia 2009a): the CNR has never really implemented project management rules; rather the whole institution hybridizes project management rules with those of the old organization in order to cope with environmental uncertainties and retrenchment. Moreover, the extended portfolio of market-oriented research projects generates high administrative burdens. In fact, according to the resource-dependence theory, PRBs need more time to administer third party funded activities, that is, the search for, and maintenance of, external funding has led to an administrative overload at all levels; Gornitzka et al. (1998) show that those contribute to academic and administrative bureaucratization.

These results come to the fore by cluster analysis applied on some key research institutes operating in the North West of Italy. Two groups of institutes are defined, A and B, that differ in the organizational behavior of their research units: Group B displays an average scientific production that is higher than that of Group A (see Table 5.3); on the other hand, Group B displays lower average values for all manifest variables that serve as bureaucratization indicators (Coccia 2009a, 2009b). In a nutshell, *in PRBs, when the time spent on administrative activities increases (i.e., more academic bureaucratization), scientific production of the research institutes decreases over time.*

The main causes of the academic bureaucratization tendency in new morphology of PRBs are due to: diseconomies of scale, generated by coordination problems of decentralized units and an administrative

Table 5.3
Descriptive Statistics of Cluster Analysis on CNR Institutes (Adapted from Coccia 2009a, 2009b)

		Arithmetic Mean		Standard Deviation	
Variable	Units	Group A	Group B	Group A	Group B
SPpI	Number of papers	3.1	3.5	1.3	1.7
T1	Months	4.8	4.9	0.9	0.1
T2	Months	5.8	2.6	1.5	2.9
T3	Days per month	2.1	0.4	1.9	0.2
T4	Hours per week	2.8	0.5	2.9	0.1
T5	Months	10.5	7.5	3.3	0.5
T6	Days	51.5	47.9	11.5	1.1
T7	Days	35.3	32.4	9.4	2.9
N	Number of documents	6.3	4.6	1.8	0.3
Institutes (No.)		9	2	9	2

SPpI: scientific production per institute; T1: contracts-staff recruitment (average time needed to recruit fixed-term contract personnel); T2: organization of events (time needed to organize events such as meetings, seminars, and projects); T3: activities in one month (time needed to participate in meetings and to draw up projects); T4: drawing up final balance sheets and budgets (time needed to compile budgets and to draw up final balance sheets of research projects); T5: approval by the headquarters (time elapsing from the project application or joint agreement/collaboration to the moment when it starts); T6: financial activities (time needed to approve budgets and to make changes to the expenditure capacity of the research unit); T7: purchases (time needed to purchase scientific material, books, journals, etc.); N: documentation (number of documents filled in).

burden at the respective headquarters; resource-dependence generates a search for—and maintenance of—third party resources and, as a consequence, administrative overloads to manage the extended portfolio of small, externally funded research projects; and an environmental complexity, caused by regulations to demonstrate accountability and to reduce public debts (Coccia 2009a, 2009b).

The effects on scientific production: Market-oriented behavior and bureaucratization (see Table 5.3) seem that affect the scientific production of public institutes negatively. The new entrepreneurial research

Table 5.4
Concentration Index R of Gini of Basic Research and Applied Activity Production within the Italian Public Research Institutes (Coccia and Rolfo 2008)

Years	Applied Activity of Research Units (%)	Basic Research Measured by	
		International Publications (%)	Total Publications (%)
2000	62.6	48.8	43.0
2001	70.9	48.9	40.0
2002	71.1	47.1	40.6
2003	69.5	45.9	37.5

organization, focusing on technological services, tends to diminish research output (as measured by bibliometric indicators): researchers tend to dedicate an increasing part of their time on services and administrative matters (e.g., preparing grant or project applications, managing grants or projects, etc.).[7] In fact, an analysis of all institutes of the National Research Council of Italy shows that there is a growing concentration of institutes with a focus on applied—fund generating—activities: Gini's concentration index (R) increased from 62.6 percent in 2000 to 69.5 percent in 2003. On the other hand, the concentration of institutes with a focus on basic research decreased over time, from 48.8 to 45.9 percent and this trend appears ongoing (see Table 5.4).

To cope with a changing research landscape, scientists tend to follow three strategies to secure external funds: the targeting of easy resources; the targeting of all available resources; or the targeting of appropriate sources (Laudel 2006). Moreover, to increase the likelihood of third party funding, scientists and research groups change the content and the focus of their research in that they: "select externally predetermined topics; diversify their research; avoid risky or hot research topics" (Laudel 2006); tend to render all or most services and consultancies demanded by clients. Some researchers, on the other hand, have changed their "researches in a more theoretical direction, because of the difficulties in securing research grants. Excellent scientists seem to feel the effects of a forced adaptation more strongly, because they have research programs that they are dedicated to and want to focus on, and they appear less willing to compromise" (Laudel 2006; see also Gläser et al. 2002; Musselin 2007). In contrast, other researchers spend excessive amounts of time on the provision of services of all kinds, to prepare applications for—and manage—grants, and reduce in this way the time available for fundamental research, with negative effects regarding the productivity of public research institutes (cf. Coccia 2001, 2004, 2008a, 2008b; Coccia and Rolfo 2007, 2008, 2009, 2010).

New Public Management Reforms: Discussion and Conclusions

Because of a volatile environment, most public R&D organizations and universities since the 1990s have tried to redefine themselves, transforming their underlying structures and developing new cooperative and competitive strategies. Moreover, continuous retrenchment of public research funds has had the impact to change the approach of many scientists towards research, with an added emphasis on technological services and applied projects directed at external actors (cf. Coccia and Rolfo 2010). Like all social bodies, the scientific profession is a living entity, responsive and adapting to external conditions that have always been in the process of flux.

As part of this change, brought about by reforms inspired by the *credo* of New Public Management (NPM), most research organizations in France, Italy, etc., have shifted from team-based organizations, structured according to scientific fields or technological competencies, to project-based organizational structures focusing on specific sets of tasks to solve problems frequently formulated extrinsically (Mihm et al. 2003). In the Italian case, the National Research Council (CNR) has never really implemented project management rules; rather the whole institution systemically hybridizes project management rules with the old organization to cope with the recent environmental threats posed by budgetary uncertainties and a shrinking public support for research laboratories. Comparing the principles and the actual practice of management underlines CNR's ambiguity in implementing project management rules and in general reform concerning the governance. The effects of this practice show up in a high organizational instability detrimental to aim and mission of the CNR. The limiting sustainability of this organization and its adopted procedures systematically lead to a managerial and organizational crisis which demands excessive amounts of human energy and financial resources.

In this context, the organizational behavior and the governance of public research is slowly turning laboratories into market-oriented institutions, focusing on applied and technological services rather than on basic research. This evolutionary process can be seen as a response to new—exogenous—exigencies and as an organizational effort to achieve legitimacy through an institutional adaptation to changes in the research environment and societal pressures. The change in the strategic orientation and mission of public research units is observable not only in Italy but in several other countries as well, for example, in France (Mangematin et al. 2006; Lepori and Larédo 2007), Germany, Australia (Laudel 2006), and Norway (Gulbrandsen and Smeby 2005).

In consideration of the widely shared objective to improve the scientific base in industrialized countries such as Italy, but above all in view of the necessity of economic systems to profit from applied research of public institutions, the organizational reforms were poorly designed without improving the inner organizational process and governance, creating confusion concerning the activities carried out within research units and by researchers. The CNR research units observe that there is an unfulfilled—perhaps even unfulfillable—role of departments in the acquisition of large projects, and they are called upon to secure third party contracts (of various kinds) to compensate for the decreasing governmental funds. This compensation is to take place in a form to assure the scientific quality in the long run, but at present, both the financial and the quality targets appear to be beyond reach. In fact, external funds, coming as they do mainly from industrial sources, are limited in extent and essentially concern consulting activities, that is, fee-for-technological services with short-term funding periods (cf. Coccia and Rolfo 2007, 2008, 2009, 2010).

The recent organizational orientation of research bodies, designed to stimulate technology transfer, technological services of various kinds, and the commercialization of research, is in line with the so-called "academic capitalism" (Slaughter and Leslie 1997) which is also characteristic of universities (Etzkowitz 2003). This strategic reorientation of research units can be seen as a functional adaptation to a growing culture of retrenchment. As a consequence, public research units adopt a market orientation similar to "quasi-business firms, with many characteristics of business firm, except for the profit motive" (Viale and Etzkowitz 2004); but they also generate higher levels of bureaucratization to manage their new activities which are based on a mix of small research projects and consulting activities for external clients (cf. Coccia 2009a, 2009b). This phenomenon has been criticized because "the embracement of the market is compromising scientific norms and commercialization (or commodification, or marketization) is in profound conflict with the function and main mission of research units and universities" (Musselin 2007; cf. also Greenfeld, 2001), that is, knowledge creation through research and dissemination through publication and education (Schuetze 2007; Slaughter and Leslie 1997). Balachandran and Balachandran (2009) claim that universities have now commodified education by the creation of powerful managerial infrastructures that have replaced academic structures. Washburn (2005) offers a highly critical assessment of close university/research units/firms ties, showing "the great and dangerous influences that money and corporate ties impose." She argues that "business and commercial interests are influencing research units and universities in an unsavory manner." Nelson (2005) states that "there are real dangers that unless [marketization of the scientific research] is halted soon, important portions of future scientific knowledge will be private property and fall outside the public domain [and] that could be bad news for future progress of science and for technological progress." Goldfarb (2008) confirms that "the growing share of research funded by industry has sparked concerns that researchers will sacrifice their scientific activity to pursue commercial goals." He finds that "researchers who maintain a relationship with the sponsor of their research experience a decrease in publications in leading international journals, implying [the danger] that academics' careers may become a function of their funding rather than their talent." Moreover, he states that "scientific merit does not necessarily serve as a funding criterion for sponsors, and citation and publication measures of scientific output are often not useful proxies for short-term social value."

The origin of this situation in Italy is the lack of a long-term national research strategy and the absence of a consistent research policy shared by the various governmental coalitions. This lack of vision and persistency in designing and pursuing policies has led to structural deficiencies, weakened the national system of innovation, and strained the overall economy as measured by GDP growth rates, productivity indicators, and measures of competitiveness and long-run economic growth (cf. Coccia 2005a; Greenfeld 2001). The hastily implemented and erratic Italian public management reforms have created problems within research organizations, particularly due to a "massification"—and consequential dilution—of academic research: "business and commercial interests have come to influence research units and universities in an unsavory manner" (Musselin 2007; Schuetze 2007).

Although this analysis focused on Italian case, and the study findings cannot be simply transferred to other research environments, the worldwide trend in the respective research sectors seems to run in parallel in many countries. This is so because universities and public research bodies exhibit a similar organizational behavior. The risk of this trend, according to Laudel (2006), is that basic research and knowledge might suffer: public research organizations have a growing tendency to focus on applied research and technological services, rather than on basic research. Certain lines of research, whose success is difficult to predict, might become "endangered species" (Laudel 2006). Such forebodings are relevant to modern, knowledge-driven economies in their support strategies for sensible changes regarding public research units and universities. Governments have to be aware of the risks associated with "entrepreneurial" research institutions, brought about by hasty reforms, because they might negatively influence knowledge production and technological trends. No doubt, further and extended researches concerning the organizational behavior of public research institutions and universities are needed, a kind of meta-research, and ways have to be found to foster, not hinder, the relevant academic institutions that play a driving role as "engines of growth," based on their intangible capital, brainpower.

Notes

1. "The national system of innovation (NSI) refers to the complex network of agents, policies, and situations supporting the process of technical advance in an economy" (Lundvall 1992). The narrow definition of a NSI would include the subsystem "research sector" represented by universities and research laboratories, while the broad NSI includes many subsystems such as finance, government, and so on.
2. A principal investigator (PI) is the lead scientist for a particular well-defined science project, such as an astronomical observatory campaign, a laboratory study, or a clinical test. In the context of federal funding from agencies such as the National Aeronautics and Space Administration (NASA) or the National Science Foundation (NSF), the PI is the person who takes responsibility for the completion of a funded project, directing research and reporting directly to the funding agency. For small projects (which involve one to five people) the PI is typically the person who conceived the investigation, but in the case of larger projects (such as the construction of scientific spacecrafts or observatories) the PI may be appointed. In the

context of a clinical study, a PI may be an academic working with grants from the National Institutes of Health (NIH) or other funding agencies, or he or she may be effectively a contractor for a pharmaceutical company testing the safety and efficacy of new medicines.

3. The Federal Government and the *Länder* (states of Germany) contribute around 47 percent of the total each.
4. Ben-David (1991) had argued that the German university of the nineteenth century prospered because it was in a position to expand, and that the expansion was properly funded.
5. See also Coccia (2008b) and Coccia and Rolfo (2007, 2008, 2009, 2010) for more details.
6. "NLAT defines a 'project' as an administrative unit, with scientific and technological objectives, a starting point and an accounting name to put in financial and activity reports. However, the same word 'project' covers two different realities: specific projects (targeted scientific or technological work in a problem-solving perspective) and thematic/generic projects, aimed at identifying, acquiring and accumulating competences in particular scientific and technological fields. A thematic project may include, or be nurtured by, an agglomeration of specific projects" (Mangematin et al. 2006).
7. The findings contained in Table 5.3 are based on data gathered through interviews of about 100 research personnel, operating in CNR-institutes located in two industrialized regions of North-West Italy. Putting research output (measured by SPpI) against administrative activities (measured by T3), one arrives at a correlation coefficient of $r = -0.20$ (sig. 0.087), and the correlation of SPpI and T4 has a coefficient of $r = -0.21$ (sig. 0.083), both indicating a negative association between scientific production and bureaucratization (administrative burden).

References

Balachandran, B., and Balachandran, I. 2009. "Performance Measurement and Change: Performance Indicators in UK Universities." In '*Proceedings*', *Performance Measurement Association,* edited by PMA. Dunedin, New Zealand.

Bartlett, A., and Ghoshal, S. "Matrix Management: Not a Structure, A Frame of Mind." *Harvard Business Review* (July–August 1990): 138–45.

Ben-David, J. 1972/1992. *American Higher Education: Directions Old and New.* McGraw-Hill Book Company/Transaction Publishers.

Ben-David, J. 1991. *Scientific Growth: Essays on the Social Organization and Ethos of Science.* University of California Press.

Bourdieu, P. 1988/1984. *Homo Academicus.* Stanford University Press (Les Editions de Minuit).

Bryde, D. J. "Project Management Concepts, Methods and Application." *International Journal of Operations & Production Management* 23, no. 7 (2003): 775–93.

Carayol, N., and Matt, M. "Does Research Organization Influence Academic Production? Laboratory Level Evidence from a Large European University." *Research Policy* 33, no. 8 (2004): 1081–102.

Coccia, M. "New Models for Measuring the R&D Performance and Identifying the Productivity of Public Research Institutes." *R&D Management* 31, no. 4 (2001): 453–64.

———. "A Basic Model for Evaluating R&D Performance: Theory and Application in Italy." *R&D Management* 34, no. 3 (2004): 267–80.
———. "A Scientometric Model for the Assessment of Scientific Research Performance within Public Institutes." *Scientometrics* 65, no. 3 (2005a): 307–21.
———. "A Taxonomy of Public Research Bodies: A Systemic Approach." *Prometheus* 23, no. 1 (2005b): 63–82.
———. "Measuring Scientific Performance of Public Research Units for Strategic Change." *Journal of Informetrics* 2, no. 3 (2008a): 183–94.
———. "New Organizational Behaviour of Public Research Institutions: Lessons Learned from Italian Case Study." *International Journal of Business Innovation and Research* 2, no. 4 (2008b): 402–19.
———. "Bureaucratization in Public Research Institutions." *Minerva* 74, no. 1 (2009a): 31–50.
———. "Research Performance and Bureaucratization within Public Research Labs." *Scientometrics* 79, no. 1 (2009b): 97–107.
Coccia, M., and Rolfo, S. "Technology Transfer Analysis in the Italian National Research Council." *Technovation* 22, no. 5 (2002): 291–99.
———. "How Research Policy Changes Can Affect Organization and Productivity of Public Research Institutes? Analysis within the Italian National System of Innovation." *Journal of Comparative and Policy Analysis* 9, no. 3 (2007): 215–33.
———. "Strategic Change of Public Research Units in Their Scientific Activity." *Technovation* 28, no. 8 (2008): 485–94.
———. "Project Management in Public Research Organization: Strategic Change in Complex Scenarios." *International Journal of Project Organisation and Management* 1, no. 3 (2009): 235–52.
———. "New Entrepreneurial Behaviour of Public Research Organizations: Opportunities and Threats of Technological Services Supply." *International Journal of Services Technology and Management* 13, no. 1/2 (2010): 134–51.
CNR Report. 2007. *Research Result and Budget*. Roma: CNR.
Crow, M., and Bozeman, B. 1998. *Limited by Design. R&D Laboratories in the U.S. National Innovation System*. New York: Columbia University Press.
Dresner, S. "A Tale of Two Ministers: Attempts at Reform of Research Systems in the Netherlands and the United Kingdom." *Science and Public Policy* 29, no. 3 (2002): 169–80.
Etzkowitz, H. "Research Groups as 'Quasi-Firm': The Invention of the Entrepreneurial University." *Research Policy* 32, no. 1 (2003): 109–21.
Fassero, I. 2004. *La gestione dei progetti: Tecniche di Project Management*. Torino: Politeko.
Gioia, D., and Chittipeddi, K. "Sensemaking and Sensegiving in Strategic Change Initiation." *Strategic Management Journal* 12, no. 6 (1991): 433–48.
Gläser, J., Laudel G., Hinze S., and Butler L. 2002. "Impact of Evaluation-Based Funding on the Production of Scientific Knowledge: What to Worry About, and How to Find Out." Fraunhofer Institut für System- und Innovationsforschung (ISI), Karlsruhe.
Goldfarb, B. "The Effect of Government Contracting on Academic Research: Does the Source of Funding Affect Scientific Output?" *Research Policy* 37, no. 1 (2008): 41–58.

Gornitzka, A., Svein, K., and Larsen I. M. "The Bureaucratization of Universities." *Minerva* 36, no. 1 (1998): 25–47.
Greenfeld L. 2001. *The Spirit of Capitalism: Nationalism and Economic Growth.* Cambridge, MA: Harvard University Press.
Gulbrandsen, M., and Smeby, J. "Industry Funding and University Professors' Research Performance." *Research Policy* 34, no. 6 (2005): 932–50.
Herbst, M. 2004. *Governance and Management of Research Universities: Funding and Budgeting as Instruments of Change.* Centre d'études de la science et de la technologie, (CEST 2004/4): www.cest.ch/en/publikationen/cestreihe.htm.
Herbst, M. 2007. *Financing Public Universities: The Case of Performance Funding.* Dordrecht: Springer.
Johnston, R., "Effects of Resource Concentration on Research Performance." *Higher Education* 28, no. 1 (1993): 25–37.
———. "Research Impact Quantification." *Scientometrics* 34, no. 3 (1995): 415–26.
Larédo, P., and Mustar, P. "Public Sector Research: A Growing Role in Innovation Systems." *Minerva* 42, no. 1 (2004): 11–27.
Laudel, G. "The Art of Getting Funded: How Scientists Adapt to Their Funding Conditions." *Science and Public Policy* 33, no. 7 (2006): 489–504.
Leilich, C. 2005. "Flat or Steep Hierarchies in Research Institutes? Empirical Findings for the Max-Planck-Institutes." Institut für Arbeitsrecht und Arbeitsbeziehungen in EG (IAAEG), Trier.
Lepori, T., and Larédo, P. "Changing Patterns of Public Research Funding in France." *Science and Public Policy* 34, no. 6 (2007): 389–99.
Lundvall, B. 1992. *National Systems of Innovation.* London: Pinter Publishers.
Mangematin, V., Blanco, S., Deschamp, B., Genet, C., and Kahane, B. 2006. "Project Management: Learning by Breaking the Rule." Working Papers 2006-06, Grenoble Applied Economics Laboratory (GAEL), INRA – Université Pierre Mendès, Grenoble-France.
Mihm, J., Loch, C., and Huchzermeier, A. "Problem-Solving Oscillations in Complex Engineering Projects." *Management Science* 49, no. 6 (2003): 733–50.
Miller, R., and Lessard, D. "Understanding and Managing Risks in Large Engineering Projects." *International Journal of Project Management* 19, no. 8 (2001): 437–43.
Mintzberg, H. 1983. *Structures in Fives: Designing Effective Organizations.* Englewood Cliffs, NJ: Prentice Hall.
Mullins, L. 1999. *Management and Organizational Behaviour.* London: Financial Times Management.
Musselin, C. 2007. "The Transformation of Academic Work: Acts and Analysis." Center for Studies in Higher Education, No. 4, University of California, Berkeley.
Nelson, R., ed. 2005. *The Limits of Market Organization.* Russel Sage.
Schuetze, H. "Research Universities and the Spectre of Academic Capitalism." *Minerva* 45, no. 4 (2007): 435–43.
Slaughter, S., and Leslie, L. 1997. *Academic Capitalism: Politics, Policies, and the Entrepreneurial University.* The Johns Hopkins University Press.
Sydow, J., Lindkvist, L., and De Fillippi, R. "Project-Based Organizations, Embeddedness and Repositories of Knowledge." *Organization Studies* 25, no. 9 (2004): 1475–90.

Viale, R., and Etzkowitz, H. 2004. "Third Academic Revolution: Polyvalent Knowledge; The DNA of the Triple Helix." www.triplehelix5.com.

Washburn, J. 2005. *University Inc.: The Corporate Corruption of Higher Education.* Basic Books.

Zeller, C. "Project Teams as Means of Restructuring Research and Development in the Pharmaceutical Industry." *Regional Studies* 36, no. 3 (2002): 275–89.

6

Centers of Learning Reconsidered in the Japanese Context

Chikako Takeishi

The objective of this paper is to examine the recent trends in Japanese university reforms in terms of the role of the university articulated in *Centers of Learning* by Joseph Ben-David (1977). The book was widely read in Japan when the translated version appeared in 1982 (Ben-David 1982). Ben-David's arguments were taken seriously in Japan, though the shortcomings of Japanese universities he pointed out in the book were too deeply ingrained in the Japanese academe then for this attention to be immediately reflected in reality. I will explore what elements of Ben-David's suggestions have actually been implemented in the Japanese academe a quarter of a century after the appearance of the Japanese translation.

In addition, I would like to argue that, while some of Ben-David's ideas became largely taken for granted in the thinking about higher education in Japan, very important points in his sophisticated argument were neglected. In fact, only the elements of the argument which could be easily interpreted in the prevailing framework of thought have been understood in simplified dichotomies, and, in time, the book was taken to represent "pragmatism" and the opposition to liberal education, philosophy, and moral education. I re-address the problems which Ben-David addressed in Japan in 1970, and discuss how his arguments remain relevant to the thinking about Japanese higher education. I argue that Ben-David's insights into anomie among university students and into other matters are even more relevant to the thinking about the problems faced by today's universities in Japan, and perhaps elsewhere.

Ben-David's Influence on Japanese Higher Education

The world-renowned sociologist, Joseph Ben-David, certainly influenced higher education all over the world through his books, but in the case of Japan, his influence went well beyond the level elsewhere. His influence on the Japanese system of higher education was very real and substantial, because he was one of the five select members of the OECD Examining Group dispatched to Japan to examine its national policies in education in 1970. Ben-David influenced policy-makers, among others, at the time when the seeds for the reform of 2004 were planted, which was the most comprehensive higher education reform in Japan in more than 100 years.

Among the changes made in the reform of 2004, the biggest change of all was turning the national universities into semi-autonomous entities called "the national university corporations." This plan had been proposed by the Central Council for Education in Japan, when Ben-David visited Japan as an OECD Examiner, and he took part in the discussion about this and other issues around reform plans for bettering Japanese higher education. After three decades and despite all the opposition and inertia, this plan was realized in 2004.

Ben-David's influence on Japanese higher education has not been limited to that of just one of the official Examiners who stayed in Japan from January 11 through January 24 in 1970. He seemed to have a personal impact on the actual carriers of the state-initiated reforms, Japanese bureaucrats. Those who directly dealt with the OECD Examiners and their Report were Japanese bureaucrats—Ben-David himself commented on the enormous difficulties the Group encountered in their attempt to communicate with interest groups other than bureaucrats during the investigation—and, the top and foremost example of these bureaucrats was Isao Amagi, the then Vice-Minister of Education.

When Ben-David's *Centers of Learning* was published in 1977, it was Amagi who translated it into Japanese. The translated version of *Centers of Learning*, which came to be widely read by academics and non-academics alike, was published in Japan in 1982 under the title of *Gakumon no Fu*, literally translated as "The seat of learning." The bureaucrats' admiration and adherence to Ben-David's basic stance was eventually embedded in the state-led programming of the reform plans thereafter.

The OECD Examiner's Report, *Reviews of National Policies for Education: Japan*, was published in 1971 (OECD 1971), and its translated version with the title, *Nihon no Kyouiku Seisaku* (Education Policies of Japan) was published in Japan in 1976. This book was translated by Junrou Fukashiro, a well-known journalist. It is interesting to note that whereas many foreign books were translated by Japanese academics, these two books were translated by non-academics.

Under these circumstances, Ben-David's ideas and sentiments had a deep influence in the making of the present style of thought concerning higher education in Japan. It was influence not in the sense of political implementation, but rather in the sense of forming the basic style of thought: promoting practical, "professional" training and science in higher education. If Ben-David could see the actual results of such influence, however, I believe he would be seriously frustrated. For there is a fundamental difference in approach between Ben-David himself and his Japanese followers.

Ben-David's arguments in the Examiners' Report of 1971 struck Fukashiro as "idealistic," according to his introduction to the translation to the report. In distinction, Amagi, the top bureaucrat in charge of education at that time, had to be realistic about the Japanese situation. There had been an irreconcilable gap between their arguments from the beginning, the gap that in fact called for the "confrontation" of the two sides. The Examining Group, including Ben-David, on the one hand, and the Japanese Delegation headed by Amagi, on the other, held a "confrontation meeting" on November 19 and 20, 1970. The basic disagreement between the Examining Group and the Japanese Delegation was recorded as Part One of the Report: it was a disagreement that reflected the fundamental difference in the views of humanity.

However sympathetic he was to Ben-David's ideals, Amagi translated his ideas and sentiments into the terms of his own realistic framework. This framework was put forward in Amagi's introduction to the Japanese translation of *Centers of Learning*, in the form of three simplified dichotomies. Amagi stated: (1) ideologically, Ben-David's position represented pragmatism as opposed to idealism; (2) regarding the content of higher education, Ben-David put emphasis on professional education as opposed to liberal arts education; and (3) methodologically, Ben-David's was a sociological approach as opposed to philosophical approach.

The recent trends in university reform in Japan are consistent with these three notions: pragmatic utilization of universities, the promotion of professional training, and the promotion of science. The actual reforms in Japanese higher education around and after 2004 include the following policies:

- Turning formerly national universities into "autonomous" entities;
- replacing fixed and secure funding by competitive grants;
- promoting university–industry collaborations;
- demanding quality assessment and certified evaluations for universities; and
- expanding graduate degrees and professional schools.

In the next section, I would like to examine the original thought and humanistic concerns of Ben-David behind his proposals to Japanese higher education in the 1970s. After examining four points on which Ben-David seemed to have particularly strong opinions, I would like to analyze the latest OECD report on Japanese higher education to see how his ideals have been realized by the recent reforms.

Ben-David's Original Proposals for Japanese Higher Education

In briefly delineating Ben-David's ideas on the role of higher education, I limit myself to the following four points which I believe are relevant in thinking about the present problems in Japanese higher education. I will present the four points in the order of the chapters in *Centers of Learning*. The first point is related to professional training, the second to general higher education, the third to research, and the fourth to social criticism.

Point 1: The Ideal of Organic Solidarity among Professionals

As for the role of the university in professional education, Ben-David stated that, in history, *old* universities functioned on "the assumption that higher education is a means for the selection and education of an intellectual elite" (Ben-David 1977, 73). That this was exactly the assumption held by the Japanese Delegation was the main concern of Ben-David and other Examiners of Japanese education in 1970, and the Japanese education system was in fact, in the words of one delegation member, "a system wholly geared to social selection functions" (OECD 1971, 26).

Ben-David envisioned a diversified society where individuals are trained into professionals and are dependent on each other in a developed system of division of labor. In applying this vision to the Japanese case, Ben-David and other OECD Examiners used the terms "autonomy" or "diversity." Ben-David's arguments about autonomy or diversity revolved around this vision of what I may term *organic solidarity* (Durkheim 1992/1893). The Japanese Delegation, in turn, did not share this vision of society, but based their arguments on their *own* notions of "autonomy" or "diversity."

Ben-David was the main questioner when the Examiners dealt with "the Central Council's proposal to turn the universities into independent corporations and to encourage university autonomy in a variety of ways" (OECD 1971, 15). Discussing this issue, Ben-David expressed his concern about the widely held social image of "a single hierarchy of prestige" in Japanese society (OECD 1971, 16). In societies based on organic solidarity as envisioned by Ben-David, diverse people in diverse professions maintain "parity of esteem," which he said he doubted whether any society with higher education system like that of Japan could maintain (OECD 1971, 26). The "social selection" role of the university combined with a single scale of excellence was incompatible with his image of "diversity." He suggested that the "diversity" pursued in the Japanese reform plan was the diversity in levels, whereas the "diversity" he proposed was the diversity "in kind." He continued, "[a]lthough Japan was more homogeneous than many modern countries, it still was not as homogeneous as it had been a hundred years previously and its present intensive division of labour demanded an educational diversity as yet not in existence" (OECD 1971, 20).

Secondly, "Professor Ben-David expressed surprise that the suggestion for reform had had to come from the Central Council rather than from the universities themselves" (OECD 1971, 16). For him, autonomy had to be attained only through active and spontaneous efforts by the actors in pursuit of that autonomy. "Professor Ben-David asked whether the Japanese Government was prepared to go that far" as to allow autonomous fund allocation by departmental self-governing units and professional university administrators of a more developed kind (OECD 1971, 16). In fact, the way in which Japanese government accomplished the 2004 university reforms from above was abrupt and had more to do with the national financial reforms than with the past arguments on educational reforms (Amano 2006, 103). The national universities in Japan put up a strong resistance to the government initiatives. The way in which reforms were carried out was far from Ben-David's original vision expressed in the term "autonomy."

In addition to the emphasis on "autonomy," the strong emphasis was also put on individuals' ability to act in the framework of democracy, that is, to respect and take account of the views of others.

> [D]iversity in universities and in forms of excellence were seen [by the Japanese] as the means of avoiding hierarchical prestige grading. University autonomy was seen as a means of achieving diversity. It is clear, however, that autonomy—the autonomy of the individual citizen to think and speak for himself while acting within a democratic framework in such a way as to respect and take account of the view of others—in itself a value, and diversity can be seen primarily as a manifestation of autonomy (OECD 1971, 36).

I will touch upon the lack in communication and decision-making abilities among Japanese participants in the discussion as pointed out by Ben-David in the section of Point 4. In the next section, I will address the value question, for that was another point of confrontation between "realistic" Japanese delegations and "idealistic" OECD Examiners, who explicitly put forward their value judgments in the quotation above by stating "autonomy [is . . .] in itself a value."

Point 2: The Role of the University in Identity Formation of Students in the Age of Diversity

Ben-David classified university students into three categories: (1) those who come to prepare for graduate studies in a specific field, (2) those who want to get acquainted with a variety of specialized endeavors and to choose from them one for graduate study, and (3) those who seek to further their intellectual and psychological maturation and to "find themselves" (Ben-David 1977, 84). For the latter two, courses relevant for professional training are not enough, because they need guidance not only in selecting their careers but also in tackling with moral questions, both of which "involve value judgments" (ibid.). With the expansion of these latter two kinds of students, the need for good general education was expanding, according to Ben-David. In reality, however, as the value of the autonomy of science spread in Japan, so did the reluctance in providing moral guidance to students. According to Ben-David, the problem of this sort was caused by a misinterpretation of the principle of value neutrality (Ben-David 1977, 169).

Ben-David proposed that "devoting serious attention to the problem of moral education at the universities may not only help students and the universities, but also society" (Ben-David 1977, 173). The need for a general higher education, Ben-David wrote, at a time when about half of the eighteen-year-olds enter college was greater than ever (Ben-David 1977, 87). The solution he suggested was to make general education "a pluralistic enterprise," as opposed to centralized programs, let alone uniform political indoctrination. Ben-David argued, "[i]f the programs are contrived uniformly according to some central directive, they are bound to be ineffective" (Ben-David 1977, 91).

Point 3: Managing the Support for Two Kinds of Research

According to Ben-David, support for the two kinds of science has to be based on distinct criteria. Applied research, on the one hand, should be evaluated for its applicability by its actual, mostly nonscientist, users. In terms of pure research, on the other, the investigator should decide autonomously what to investigate, and research should be evaluated by peers. Support, or funding, of such research cannot be based on possible benefits to the economy or to the solution of a practical problem (Ben-David 1977, 176). The dominant complaints about competitive research funding is related to the fact that far more weight is put on the former kind of research in Japan, the situation which leaves little room for pure, or basic, research.

The reforms to promote competitive funding have had a big impact on natural sciences in particular. The issues concerning competitive funding and other reforms in Japan were discussed in *Nature*.[1]

Administrative management in Japanese universities is almost solely limited to the question of money, but, in fact, the shortage of time seems to be a more serious problem for many academics in Japan. The lack of time management and the sloppy research product as a result were already pointed out by the Examiners forty years ago:

> The Examiners are impressed by the broad perspective thus laid down by the Central Council's report, by the high quality of the statistical background research done for the Council and by its desire to seek widespread public discussion of its proposals. However, it should be noted that the eight days allotted for the public hearings was a very short time, and the members of the Central Council's Committee's relevant sub-committees are all busy men, and not recruited on the understanding that the work should require from them a major share of their working time (OECD 1971, 72f).

Then the Report goes on explaining the situation in which five smaller *ad hoc* committees met only thirty two times each on average, and there were only five public hearings to which all universities were invited. "The research work consisted almost exclusively of a re-working of statistics already available and only one special survey was conducted" (OECD 1971, 73). Yet this situation is that of forty years ago, and among Japanese academics now, there is an almost unanimous agreement that universities today have become much busier places than ever. Today, sleep deprivation for the sake of doing research is a luxury.

Japanese professors are running around to do chores, and their lives are full of work that has nothing to do with the professional training they received.

The source of the problem is probably the persisting antagonism and distrust between groups that could be expected to cooperate. There are "professionals" outside universities who are supposed to be trained professionally for important but not academic work. Because of the mutual distrust, university professors are reluctant to contract their chores to outside institutions. This distrust from the side of the university is not totally unfounded. The distrust is related to such qualities as professional ethics and the sense of responsibility for the client organization, both of which are the necessary qualities for professionals. The hostility toward universities from outside professionals is not without foundation, either—it is grounded in their own past experiences as university students.

This problem of antagonism in Japan was also addressed by Ben-David under the title of "communication between different social groups," the issue which I will touch on below.

Point 4: Communication between Different Social Groups

The main questioner in the Confrontation Meeting on the issue entitled "Communication between different social groups" was, again, Ben-David. He introduced the question:

> [T]he Examiners' most pervasive experience during their visit was of an extraordinary blockage in communication. Thus, for example, the Examiners found that the Ministry of Education is as reform-minded and as willing to benefit from any good idea as any similar Ministry in the world. And they found that professors here have probably as much feeling of responsibility towards the junior academic staff and the students as other professors. Yet they also met almost without exception teachers who feel that the Ministry is an immovable block

> of power, which would never listen to the views and ideas of the school teachers, and there are expressed similar feelings among the junior staff towards their seniors, or among the students towards the university authority as a whole. While these communication faults usually appear between what are viewed as different levels of authority, they also are reinforced by isolation between relatively parallel groups [. . .] (OECD 1971, 32)

Ben-David in this introductory speech stressed that today the educational problems in Japan, like elsewhere, all revolved around the central issue of the preservation of democracy. And he asked the Japanese Delegation to "consider, in particular, the extent to which education could contribute to the improvement of social relationships and communication and to the support of the principle of 'co-operation between equals'" (OECD 1971, 33). Apparently, Ben-David considered cultivating communication skills for democracy to be an important role of the university in our time. Ben-David concluded the discussion:

> Education is so complex a process that central issues frequently get lost in the discussion of the complicated techniques of imparting knowledge. In such a technically outstanding system as one finds in Japan, said Professor Ben-David, this was particularly the case; technical skill in imparting knowledge tended to take priority in educational policy over the effects of the system on the pupils as human beings. [. . .] The Examiners felt very strongly that [. . .] pupils [should be able to] have a greater opportunity to interact with their peers [. . .] to acquire the skills of communication and collective decision-making. Though this would have no immediate effect, in the long run it would help future generations of adults to cope with the problems of social communication which at present seemed to be so intractable (OECD 1971, 34).

He suggested the promotion of learning "procedures," that is, of methods of disagreement in which points of conflict could be "specified" rather than "generalized." This was a practical proposal by Ben-David to solve the problems of antagonism, still present as causes of inertia and inefficiency of the Japanese society.

The Latest OECD Assessment on Japanese Higher Education after the Reform of 2004

Now I would like to turn to *OECD Reviews of Tertiary Education: Japan (The Reviews 2009)*. Its translated version in Japanese is entitled

Nihon no Daigaku Kaikaku, meaning "University Reforms in Japan," because the investigation for this report was conducted from 2004 to 2006, right after the higher education reform of 2004.

The first point was about organic solidarity among professionals. Has the Japanese higher education system ceased to be the social selection system? *The Reviews 2009* evaluates the "diversity" of Japanese higher education system as follows: "The mixture of universities and colleges, public and private, national and local, add up to a system of considerable diversity. This is a characteristic that we see as a notable strength of the Japanese system" (Newby et al. 2009, 25; OECD 2009, 34). It, however, states that this diversity also functions as a social selection system in Japan:

> The sector is not only diverse; it is also mature. In Japan there is a wide public consensus on the status hierarchy of institutions, both between and within each category. Thus while the sector is nominally very diverse, in reality it is very hierarchical. Each institution "knows its place" and there is a societal consensus on institutional rankings. Statistically Japanese tertiary education is a 'mass system'; but culturally it retains all the attributes of an 'elite' model. Consequently there is fierce competition to enter the most prestigious universities, public or private. There is a widespread adoption of supplementary private instruction of those students seeking entry to the 'top' universities and a number of prestigious private universities are also vertically integrated into the schools system through having established their own (private) feeder schools. Therefore the universities continue to perform one of their traditional roles in Japanese society, providing a kind of social sorting mechanism for entry into the elite professions in both the private and the public sectors (Newby et al. 2009, 26; OECD 2009, 35f).

The diversity within the Japanese higher education system remains to be largely the diversity in levels, as opposed to Ben-David's vision of the diversity "in kind." Still today, universities in Japan stand aloof from the rest of the society. *The Reviews 2009* states:

> Thus, for example, a survey of higher education graduates in Japan and ten European nations conducted at the end of the 1990's found that 54% of European graduates reported making 'extensive use of knowledge and skills acquired during [university] study'—as compared to 24% of Japanese graduates. While European and Japanese graduates were equally likely to report that problem-solving was a key competency for working life, 58% of European students reported that this was a competency acquired by the time of graduation—as

opposed to 39% of Japanese graduates (Newby et al. 2009, 63; OECD 2009, 80).

Universities in Japan, in general, are not responding to the needs for professionalization of society at large. This situation in Japan is responsible for the unprofessionalism of the center of learning itself. According to the OECD report, the Japanese society does not yet have a pool of management professionals in university administrations (Newby et al. 2009, 20; OECD 2009, 28). Even researchers are not trained professionally because of "a lack of 'practical' education in research planning and management, experiments, and data processing" (Newby et al. 2009, 76; OECD 2009, 96).

The second point was about the role of higher education to provide pluralistic general education in societies with diverse identities and lifestyles. According to the observation in the *OECD Reviews 2009*, the reformers of higher education are unable to define the goals, and thus are just following some strategic guidelines. According to *The Reviews 2009*, the main weakness of Japanese higher education reforms is the lack of a concrete vision. Instead of concrete visions, the Japanese reform plans include only abstract notions or imperatives such as "increased institutional differentiation" or "differentiate":

> Planning documents [by the Ministry of Education, Culture, Sports, Science, and Technology] contain injunctions to institutions ('differentiate') but they do not yet provide a vision of a tertiary system, outlining how much diversity, of what sorts, and to what ends, is desired. There is an emerging repertoire of policy instruments that hold the potential to connect ends and means, such as the 21st Century Centers of Excellence Program, but some are not clearly connected to a wider vision of the tertiary system (Newby et al. 2009, 20; OECD 2009, 28).

The social environment which surrounds Japanese higher education is in a state of anomie.

The higher education itself is responsible for the anomic situation at large. The importance of the identity-formation function of university has been underestimated in Japan, after the indoctrinating education during World War II and a misinterpretation of value neutral science ever since.

The "professional" education, which in the context of Japanese higher education means the exam preparation training for national certifications such as National Bar Examination or Certified Public

Accountant, has been strongly promoted in recent years. *The Reviews 2009* refers to this trend:

> Universities have also developed add-on courses and modules, outside the normal course of study, which have the purpose of providing students with work related skills. These may be offered either by universities, or *senmon gakko* (vocational schools) set up alongside their campuses. Other universities have tried to increase their attractiveness to students by facilitating 'double schooling' on campus either by hiring in lecturers from specialized schools or by contracting with specialized schools to offer 'on campus' programs (Newby et al. 2009, 67; OECD 2009, 84).

In addition to professional education, it is important for higher education to provide general education in a pluralistic form that enables people to participate in shaping their society. If the emphasis on "professional" education leads to downplaying general education, it might result in producing not only "elites" without a social vision but also a mass of people suffering from confusion of identity, which will weaken motivation, impede the integrity of will, and eventually endanger the sanity of one's mind. Research on the influence of anomie on human minds is still under way, but we should not underestimate the serious social consequence of anomie and the importance of general education, which Ben-David emphasized so much.

Thirdly, what was the impact of reforming research funding system in Japan? The financial pressure on basic research is certainly the downside of the reform. The system of competitive funding was introduced in the country where public spending on higher education and research is relatively small compared to other OECD countries. The OECD reviews of 1971 revealed that the national funding of higher education in Japan is the lowest among OECD countries. The situation is the same in 2009. *The Reviews 2009* states:

> Total public expenditure on tertiary education institutions in 2003 equaled 0.6% of GDP, as compared to the OECD average of 1.3%. Japan stands alongside Korea at the bottom of OECD member countries in public tertiary expenditure, and below countries such as the United States (1.5%), Canada (1.7%), and Australia (1.1%) (Newby et al. 2009, 40; OECD 2009, 53).

There are, however, some upsides to the reform, too. The increase in university–industry collaborations, for instance, is positively assessed by *The Reviews 2009* (Newby et al. 2009, 74f; OECD 2009, 93f).

In addition to scarcity of money, scarcity of time is perhaps a more pervasive problem for researchers in higher education. *The Reviews 2009* sets the framework to distinguish the old-fashioned model of Japanese university system, which *The Reviews 2009* calls "the Humboldtian model of the university" and the universities that are "more agile, more responsive to societal needs, more innovative, creative and enterprising, and more competitive, in terms of both teaching and research quality" (Newby et al. 2009, 18; OECD 2009, 25):

> For example, a 2005 survey of undergraduate students in eight Japanese universities revealed that nearly half of students (47%) reported that they receive no 'advice or guidance about their educational program' from their professors [Yamada, 2007]. Japan's university system was deeply influenced by the German university tradition, and the Humboldtian model of the university. For many professors, their status and reputation have been rooted in their accomplishments as a researchers and members of an expert community, rather than as an undergraduate teacher. Moreover, the circulation of personnel between universities and the working world is modest by international standards, (e.g. between universities and firm-based research labs), strengthening the separation of tertiary education, particularly university education, from working life. This culture of university life has shaped the wider tertiary system, and even instructors in junior colleges noted their obligations as scholars—an orientation one would not find in like institutions elsewhere in the world (Newby et al. 2009, 62f; OECD 2009, 79).

In Japanese higher education, there coexist two types of professors: those whose research style is based on the Humboldtian model, and others who give guidance to their students or practical instructions in research planning and management, experiments, and data processing, and are involved in some university–industry collaborations, prevention of "academic harassment" (referred as *akahara* in *The Reviews 2009*), quality assessment, or "Support Program for Distinctive University Education" like "Good Practice" (GP), and other practices which such outside observers like OECD positively evaluate. At the moment, because of the lack of administration that can allocate research time among professors, the system results in rewarding the former in effect. Unproductive overwork among university staff is a big problem.

Fourthly, what is the situation regarding the problem of social communication now, which was called "intractable" in *The Reviews 1971*? Perhaps it suffices to point out the following passage from *The Reviews 2009*:

> As we have already indicated, the introduction of these reforms—let alone their implementation—created great controversy within the university sector, both national and private. The Humboldtian model had bequeathed to Japanese universities an extreme form of self-governance, with control in all national, public and many private universities in the hands of a professors' council. [. . .] For the bulk of institutions [. . .], all academic decisions had rested with the professors' councils in each faculty and while financial decisions could be made by the university's board, in practice the professors' councils had huge powers of veto, without being responsible for the financial and strategic consequences of their decisions (Newby et al. 2009, 32; OECD 2009, 41).

There usually is constant antagonism and mutual distrust between the faculty and the board. There are also decades-old, unconstructive conflicts among faculty members themselves. As *The Reviews 2009* points out, it is almost impossible to arrive at positive and pro-active consensus within this "extreme form of self-governance." This "extreme form of self-governance" of university faculties, however, has sustained many Japanese professors' strong commitment to education and their organization. The sense of participation in the decision making has been the source of good-will that has enabled each faculty member to make intensive efforts in educating their students or to take extensive workloads. When a good decision-making system is combined with an administrational professionalism, it might become an excellent self-governing organizational style. At the moment, however, "acquiring the skills of communication and collective decision-making" as pointed out by Ben-David (OECD 1971, 34) is still an important task to be faced by Japanese society at large.

Conclusion

Historically speaking, the major role of modern university has been to provide diverse professional training, according to Ben-David. In order to make the efficient division of labor and collaboration among different professionals possible, the need for general higher education is much greater than ever before. Not only each professional has to have sufficient skills and knowledge for the profession but also each professional should be oriented by his or her sense of identity in career choice, and should be able to make moral and other judgments of one's own. There are many problems in Japanese higher education in these regards, and among them are the shrinking of general education despite the prevailing state of anomie among students, because of the overemphasis on professional education, misinterpretation of value neutral science, and unproductive overwork of professors. Problems of overwork, misinterpretation of moral problems, and the neglect of basic research themselves are the product of the models of the university in the past. The solutions to these long-term problems are also largely in the hands of universities themselves.

Even now, almost four decades after he formulated his views in the "confrontation meeting," Ben-David's general arguments on higher education articulated in *Centers of Learning* and his concerns about the Japanese higher education seem very relevant. And even now that the Japanese system has accomplished, however partially, the materialization of pragmatic and realistic elements of his arguments after all these years, his ideas still provide for us some practical, but long-term, solutions for the remaining problems the present Japanese universities face.

It was Ben-David's firm position that social science cannot predict the future because "[s]ocial institutions are open systems, and no one can know whether their future will be determined by the same variables that determined their past and present" (Ben-David 1977, 3). The principle of academic freedom, according to which higher learning and research have to be free from political, religious and ideological control, had not been seriously challenged, according to Ben-David. "However, all this may drastically change" (Ben-David 1977, 4)—his visits to Japan in 1967 and 1970 must have been responsible for this observation. Was the Japanese case going to be what seemed to him an external threat to the tradition of academic freedom? How about other Asian cases? He didn't know, because social institutions such as education were open systems. And the answer to that historical question is open to us. The history is always in the making.

Note

1. For example, in the issues of December 22, 1994; January 5, 1995; February 15, 1996; June 10, 1999; September 16, 1999; September 30, 1999; May 13, 2004; and June 30, 2005.

References

Amano, I. 2006. *Daigaku Kaikaku no Shakai-gaku.* Tokyo: Tamagawa University.

Ben-David, J. 1977. *Centers of Learning.* New Brunswick, NJ: Transaction Publishers.

______. 1982. *Gakumon no Fu.* Tokyo: Saimaru Shuppankai.

Durkheim, E. 1992/1893. *The Division of Labor in Society.* New York: The Free Press.

Howard Newby, Thomas Weko, David Breneman, Thomas Johanneson and Peter Maassen (2009) are quoting from: Yamada, R. 2007. "Learning Outcomes of College Students in Japan." Paper presented to 2007 Annual Meeting of the Association for Institutional Research, Kansas City. All quoted in OECD (2009).

OECD. 1971. *Reviews of National Policies for Education: Japan.* Paris.

OECD. 1976. *Nihon no Kyoiku Seisaku.* Tokyo: Asahi Shimbunsha.

OECD. 2009a. *Nihon no Daigaku Kaikaku.* Tokyo: Akashi Shoten.

OECD. 2009b. *Reviews of Tertiary Education: Japan.* http://www.oecd.org/dataoecd/44/12/42280329.pdf

7

The Autonomy of Science after Modernity

Yaron Ezrahi

Joseph Ben-David did not consider his work on the sociology of science and higher education simply an interesting subject worthy of sociological research. He was deeply engaged in it with a sense of mission. Ben-David considered autonomous science, free universities and science-based culture as the weapons of the liberal democratic order against the dangers of fascism, communist totalitarianism, and mob violence. As a young student in the early 1960s, I was both moved and convinced by Ben-David that science-based rationalism provides the only powerful basis for politics which can resist the myths idealizing the use of force to restructure the political order. But Ben-David was not a naïve rationalist. He recognized the complex fragile social and institutional bases of knowledge and rationality as forces in society and politics and attempted to discover the conditions and processes which can support and preserve the autonomy of science and higher education after World War II. In retrospect, obviously, what Ben-David regarded as conditions congenial for the autonomy of science during the decades immediately following that war could not remain the same in later decades.

Ben-David belonged to a generation of scholars-intellectuals—like Popper, Habermas, Shils, and another former teacher of mine at the Hebrew University, Zeev Sternhel—who believed that fascism was a consequence of the eclipse of the enlightenment and that the way to cope with mass irrationalism and totalitarianism is to restore respect for scientific truth, the evidence of certified empirical facts, and rational arguments. He very much approved at that time, less than two decades after World War II, what the Nobel Prize laureate Jacques Monod called "the ethics of knowledge," the ethics of a society based on the normative structure of science as an enterprise of a community of rational individual truth seekers, self-governed by a commitment to uncoerced persuasion (Monod 1971/1970). This was also the position of Albert Einstein who, in his introduction to an edition of Galileo's New Science, praised the great Italian scientist for liberating his generation from mythological thinking, a goal which Einstein thought defines the mission of science in our time as well (Rowe and Schulmann 2007, 132–34).

Today, when the concept of "objective facts," both within and outside science, is problematized by the best scientific minds of our time, as well as by historians like Mary Poovey (1998) or philosophers like John McDowell (1994), we tend to forget how crucial it was in the defense of the universities, the free press, and the freedom of expression and information in the years immediately after the war. The popular belief that science can unambiguously certify or discredit factual assertions rendered the social authority of science a valuable resource in legal or political conflict resolutions. Such expectations seemed to correspond to the liberal ethos of reason, objectivity and truth-based ethics of public affairs (Ezrahi 1990).

In the long run, however, a stand that endorses these ideas could not escape the critics who insisted that in the context of public affairs these ideas are supportive of liberalism, as a contested ideology, not a universally acceptable idea. The generations of historians, philosophers, and sociologists of science who followed the early post-war advocates of the ethics of knowledge could not ignore the complex intellectual, social, economic, technological, and political processes that have been challenging the intellectual and normative presuppositions that supported the autonomy of science and its institutional matrix in the earlier period. Today, we cannot simply believe with Ben-David that the university can—and should—provide what he called "moral guidance" to its students, and that scientists are unquestionably qualified to instruct the public how to distinguish between "reason and prejudice." Nor is it reasonable to believe that the challenge faced by higher education today, more than 30 years after Ben-David (1977a) wrote his landmark essay prepared for the Carnegie Foundation for Higher Education, is posed by the fact that, in his words, we are living in a society "in which decisions in every walk of life are increasingly based on research [. . .]". Without getting into a thorough discussion of the profound changes in the political and ideological landscape of contemporary liberalism (Ezrahi, forthcoming), I would like to note that contemporary liberal thinkers have increasingly come to realize that the weakness of liberalism in its former confrontations with fascism and communism was the (1) emotional aridity of its concept of rationality as a realistic guide for action; and (2) the poverty of liberalism's conceptions of freedom and market values—as a source of meaning in the daily life of laypersons—and liberalism's poor, or unsatisfactory, conceptual, and normative development of the relations between its concept of freedom and the idea of social justice.

Observing the clash between fascism and the liberal university in authoritarian societies, Ben-David naturally thought that the greatest threat to the autonomy of science and the freedom of the university comes from the political intervention of the illiberal state. He did not realize that precisely in the liberal-democratic state the danger to the autonomous status of science and the university would come from the very liberal institution of the market, from the voluntaristic economic-business sector or, more precisely, from the crawling penetration of market values and interests into the heart of the university and its academic practices. Although this process started already during World War II with the policy of the American Office of Naval Research of contracting out government scientific and technological functions to private firms, it has taken ominous dimensions mostly since the late 1980s, when fields of science and expertise that could accelerate their development—due to the availability of private (and public) money—did not correspond to fields of science that would have been given priority if the primary criteria for allocating resources to science would have been the degree of intellectual promise and expected contribution to the advancement of knowledge.

In order to better understand the process by which the terms of the autonomy of science were transformed since the late decades of the last century, I would like to use a version of a simple exchange model I used 30 years ago at the Harvard seminar on science and public policy when it was directed by Don K. Price, Harvey Brooks, and I. B. Cohen. It goes like this: Scientists and politicians (and government bureaucrats), respectively, control resources vital to the activities of each other. *Scientists* were believed to control:

- knowledge vital to the decisions and programs of politicians and other public servants;

- scientific or expert authority vital for legitimating public policy as instrumentally valid and, therefore, depoliticized means to promote public goals;
- intellectual and technological resources that appear to render the process of government decision-making and its results both more objective, transparent, accountable and, therefore, more amenable to outside public scrutiny (often reinforced by public regarding experts); and
- the provision of vital trained technical manpower and technologies in such crucial areas as defense, medicine, and economic policy.

Politicians, on the other hand, were believed to provide:

- material sources vital for financing of the increasingly costly instruments of basic research (linear accelerators, telescopes, etc.);
- political and symbolic resources useable for the public legitimation of the large material investment in science and the endorsement of its value to society;[1]
- the encouragement of students to study science and to commit themselves to demanding careers in mathematics, physics, chemistry, biology, etc.

I would like to argue that in the course of the second half of the twentieth century, the terms of this exchange—or social contract—between scientists and politicians have been transformed, mainly due to the decline in the political demand for university-based scientific knowledge and authority, and in the process the vacuum that was created by this decline was quickly filled by values, interests, and practices of the private business sector.

Business firms not only replaced many government agencies as clients of scientific research and development, but also became central in mediating the uses of science by the state. An important aspect of these developments manifested itself in the massive uses of the authority of experts and scientific visual props and metaphors in commercials in the mass media and the partisan engagement of scientists in the competition of business firms for niches of the market. Obviously, against such a background, the role of the university in providing moral guidance could easily and tacitly mean socializing young scientists into capitalist rather than public or civic values. In other words, the ideological alliance of science with liberal-democratic civic values was increasingly eclipsed, or replaced, by the association of science and technology with neo-liberal capitalist-market values of the private sector.

Correspondingly, the practice of secrecy with respect to the results of research and development for privately owned firms, and the priority given to marketing strategies guided by the goal of rushing to conquer a drug or an electronic market, rather than waiting for more complete and adequate results of research, were inherently incompatible with the kind of norms sociologists like Parsons, Merton, and Ben-David thought should not only govern science, but also be extended from the scientific community to the larger society.[2] The exchange between scientists and politicians was disrupted also by the fact that politicians learned very quickly to pill off the authority of experts from the bodies of knowledge in which it is grounded and deploy it as an elastic political resource barely constrained by the substance of expertise. Setting up committees of investigation, staffed by experts whose reports ended up languishing in locked drawers, became a standard practice of government officials who sought to ride a crisis—like disaster in the space program or a failed nuclear installation—until public alertness and anxiety subside. It is, of course, well known that scientific controversies around such publicly salient issues as stem cell research, genetically modified organisms, measurements of IQ and their meaning, and food additives, have further eroded the social authority of science (Ezrahi 1976). The public has continued to falsely believe that the authority of science emerges from the inherent certainty of "real factual" scientific truths and that a public controversy among scientists indicates the failure of some scientists to represent legitimate scientific knowledge.

This gap between the simplistic public perception of science and its inherent complexity could be easily exploited by interested parties to discredit scientifically certified knowledge when its implications contradicted their preferences, and endorse uncertified claims of scientific knowledge when they serve them. Because probabilistic formulations of scientific knowledge have been usually much less persuasive to laymen than unwarranted rhetorical claims of uncontestable scientific truths, politicians, entrepreneurs trying to raise money for science, and commercials could systematically ignore with impunity the actual boundaries of scientifically warranted knowledge. The deleterious impact of public controversies among scientists on the social authority of science has been augmented by the fact that, although such controversies usually develop on unresolved margins of much vaster scientific consensus, they can easily and indiscriminately discredit the social authority of entire bodies of knowledge. It is noteworthy, however, that often, like in the court of law, also in the court of public opinion, assertions of facts are selected out and edited as inputs into complex clusters of competing values and beliefs and, therefore, in both contexts evidence has been always a matter of policy no less than facts.[3]

The point I wish to emphasize is not that science has actually become less relevant to decisions and actions in the context of public affairs; rather, it has been integrated into the context of public affairs as a much more diffused factor, rarely distinct from legal, economic, political, and ethical factors and, therefore, much less delineated and salient as the culturally and psychologically charismatic factor recognized formerly as "science" which was loaded with the authority, and the promise, of constituting rational alternative to prejudice, ideology, and arbitrary politics. As I have noted above, the decline in the political attractiveness of science, and the political demand for its symbolic and intellectual resources, created a vacuum into which the private sector quickly entered with its considerable demand for both the symbolic resources of science and scientific expertise. The whole culture of private endowments to universities and scientific institutions, of course, partly reflects the transference of the remaining aura of science to private individuals and firms. Moreover, it is enough to visit *Disney Future World* in Florida and seeing the grandiose commercialized exhibitions of business corporations like General Electric, IBM, Boeing, and other technologically intensive firms to realize how such privately owned companies have inherited the distinct powers of science to represent, for the young generation, "the future" and its idealized public good. Over recent decades, this transfer has been manifest also in the pathetic attempts of NASA to revive the popular enthusiasm that accompanied the flights to the moon about four decades ago.

The aforementioned decline in the direct political-public demand for science has not been a passing development, but a result of deep changes in the fabric of democratic politics. I would like to suggest that these changes in the values and practices of contemporary politics are the most important cause of the transformation that made earlier notions of the public role of science and civic role of higher education anachronistic, despite their persistent normative power within the intellectual community and some wider social circles. Without getting into details, I would like to emphasize three related factors:

- the overall social decline of faith in the enlightenment vision of the partnership between science and democracy;

- the erosion of normative holism as a frame for systemic social and political engineering of public policies and programs; and
- the accumulative impact of the continual exposure of the lay public to the electronic mass media on its diminished confidence in its own ability to distinguish between facts and fictions.[4]

The importance of the erosion of public trust in the enlightenment utopia of rational, deliberative democracy cannot be brushed aside without profound misunderstanding of the historical role of utopias in the creation of movements and institutions. The history of politics in modern time has largely been the history of shifts between alternative unrealistic visions of ideal political orders. Science was clearly the centerpiece of the last one; it was extraordinarily important in the minds and expectations of wide social circles and in democratic political rhetoric since the eighteenth century. The reasons for its decline are too well known to be repeated here. But an important aspect of this change is what I would like to call the erosion of moral holism and the rise of moral perspectivism. By this, I mean that the notion that there is an acceptable order of values which can be a binding foundation of consensus on priorities of public allocations and actions has lost its grip in contemporary Western democracies.

Consequently, democratic politics, more visible than ever before, has become an eclectic configuration of multiple power games between individuals and groups whose consequent policies are incoherent shifty products of temporary balances between diverse normative perspectives and interests. Such a political context is basically uncongenial for the role of science in public affairs as conceived by Ben-David and his fellow members of the community of the faithful of the enlightenment.

Finally, this process has been accentuated by the popular impact of the mass media and modern cinema on the public faith in its own capacity to distinguish facts from fictions, or the credibility of expert authorities to unambiguously instruct laypersons in this matter. The controversies around "intelligent design" and naturalistic medicine are cases in point. Today, the authority to accord observations or other types of experience the status of binding facts has been seriously weakened both within science and in the wider social context of public communications. Laymen intuit today that "facts" are often relative to values, interests, and concepts, that what has been taken for a long time as a given (like the reliability of economic indicators) may be baseless, that safe food may suddenly turn to be poisonous, or that miracle drugs of the recent past may be unhelpful or even positively harmful. As far as the lay public is concerned, science has failed to provide safe unambiguous guidance that reduces anxiety and the burden of making hard choices in vital matters of life and death. The promise of science, to emancipate laypersons from the burden of the necessity and responsibility of choosing between alternative opinions or courses of action, has evaporated in many areas. In some areas, the public shifted to more satisfying alternatives ranging from naturalistic medicine to the renewed alternative authority of religion.

I would like to repeat that what I am discussing here is not so much the diminished importance of scientific knowledge to individual and civic life, as the diminished saliency of science and higher education, as means and foci of lay and political hopes that knowledge is a paramount instrument for the improvement of life and the solidification of democracy. One of the consequences of the diminishing value of science, as the cultural and political resource that enabled a degree of symmetry or balance in the assets exchanged between scientists and politicians, was to disrupt that balance and tilt it in favor of the politicians. The institutional powers of science to protect its autonomy from political—and even more so from economic—pressures have been steadily declining.

Returning to Ben-David, I conclude that his concerns for protecting the integrity of science and higher education against political interventions have been compounded now by concerns about the transformation of institutions of higher learning into service stations that train producers and consumers, rather than thinkers and citizens. At the very time processes of political fragmentation have been eroding the authority of science in legitimating coherent collective visions of the future and of legitimate public actions, the market has been already eroding the internal intellectual and institutional navigation tools of science. It may be pretentious to predict where these trends lead. But it seems obvious that the structure and terms of the autonomy of science and its relation to public affairs are undergoing in our time profound transformations.

Notes

1. The famous statement of the first man on the moon, "One small step for man, one giant leap for mankind," can serve as an illustration of the power of symbolic legitimation of the scientific enterprise.
2. For Parsons on expanding university norms to the larger society, see: Parsons and Platt (1973).

3. By policy I mean the often eclectic and not always explicit integration of political, legal, economic, ethical, and technological considerations into the formulation of alternative courses of decision and action.
4. I have discussed these factors extensively in my book: *Imagined Democracies: Necessary Political Fictions*. Cambridge-Newyork: Cambridge University Press, Forthcoming 2012.

References

Ben-David, J. 1977. *Centers of Learning: Britain, France, Germany, United States*. McGraw-Hill/Transaction Publishers.

Ezrahi, Y. 1976. "The Jensen Controversy: A Study in the Ethics and Politics of Knowledge in Democracy." In *Controversies and Decisions: The Social Sciences and Public Policy*, edited by C. Frankel. Russel Sage Foundation.

______. 1990. *The Descent of Icarus: Science and the Transformation of Contemporary Democracy*. Cambridge, MA: Harvard University Press.

McDowell, J. 1994. *Mind and World*. Cambridge, MA: Harvard University Press.

Monod, J. 1971/1970. *Chance and Necessity: An Essay on the Natural Philosophy of Modern Biology (Le Hasard et la Nécessité. Essai sur la philosophie naturelle de la biologie moderne)*. Knopf (Seuil).

Parsons, T., and G. M. Platt. 1973. *The American University*. Cambridge, MA: Harvard University Press.

Poovey, M. 1998. *A History of the Modern Fact: Problems of Knowledge in the Sciences of Wealth and Society*. Chicago, IL: University of Chicago Press.

Rowe, D. E., and R. Schulmann, ed. 2007. *Einstein on Politics: His Private Thoughts and Public Stands on Nationalism, Zionism, War, Peace, and the Bomb*. Princeton, NJ: Princeton University Press.

8

The Crucible of Human and Social Sciences: Nondisciplinary Approaches to Human Subjects and the Role of the Intellectual in Mid-Nineteenth-Century France

Nathalie Richard

This essay focuses on the emergence and social foundation of the "role of the intellectual," which allowed for a very special kind of scholarship in nineteenth-century France. In many respects, my approach can be related to Joseph Ben-David's different explanations of the emergence of science in general—and of specific disciplines (psychology and bacteriology)—as a result of "the scientist's role" and, in regard to the specific disciplines, of "role-hybridization" (Ben-David 1971; Ben-David 1960a; Ben-David and Collins 1966). Thus, this essay can serve both as an exemplification of Ben-David's theory and as an example of his relevance to the fields of human and social sciences with which he was not specifically concerned. But, dealing specifically with these fields, my approach also diverges from Ben-David's view on one point: the exclusivity of the university as the sole institution for modern science, a strong thesis in Ben-David's historical sociology (Ben-David 1972b, 1977a).

These topics will be approached from an historical point of view, using the examples of three influential intellectual figures in mid-nineteenth-century France. Two of them, Hippolyte Taine and Ernest Renan, are well known, whereas the third, Alfred Maury, has been mostly forgotten (Carroy et al. 2007). He was a specialist of the history of religion and of popular culture, a member of the French *Académie des inscriptions et belles lettres,* a professor at the *Collège de France,* where he succeeded Michelet in his chair, and a collaborator of the Emperor Napoléon III in the latter's *History of Julius Caesar*. He was also the author of a famous book on dreams (Maury 1861), which was a main inspiration for Freud's interpretation of dreams (Carroy 2006a). As the curator of the Emperor's library and of the National Archives between 1868 and 1888, he helped many writers to access books and documentation. Gustave Flaubert, for example, asked for his help when he undertook some historical research for *Salammbô* (1862). As a member of the imperial circle, Maury was also a useful intermediary to obtain a position or a grant. He played an important part in the election of his friend Ernest Renan to the *Collège de France* in 1862. Therefore, Maury was at that time an influential member of the French intellectual community. Around 1860, his name was as well known as Renan's, Taine's, or Émile Littré's.

These authors were often associated by conservative thinkers as representing a materialist and atheist trend in philosophy, which could endanger the social and moral basis of the modern world. Catholic orthodox writers, such as the Bishop of Orléans, Félix Dupanloup, or conservative spiritualist philosophers who held the main positions in French academic institutions, such as Adolphe Garnier or Elme Caro, depicted them as dangerous radicals, from whom young readers should be protected (Caro 1864; Dupanloup 1874/1863). On the opposite side, liberal and republican opponents to the conservative regime of the Second Empire, turned their names into a slogan and used their works, inspired by positivism, as a basis for their reformist political ideology. After the establishment of the Third Republic in 1870, Taine and Renan became quasi-official state philosophers. As their renown and influence became overpowering, their philosophy also came under criticism, either philosophical, as in the case of Henri Bergson, or more ideological and political, from the novelist Paul Bourget and the essayist Maurice Barrès, for example.

Such was the intellectual context in France at the end of the nineteenth century when the modern academic fields were established in the new university system.[1] In their vast majority, those who are now considered as the founders of the modern human and social sciences were educated in "humanities" (*les lettres*), a field in which philosophy occupied a dominant position (Brooks 1998) and which was characterized by the blurring of the lines which now distinguish disciplines in the human and social sciences. In their curricula, Taine's and Renan's works were compulsory readings, setting up a model for the science (or sciences) of man. The founders of the modern human and social sciences, therefore, made strategic reference to their works. They defined a new scientific status and claimed autonomy for psychology, sociology, geography, history or philosophy, both against and in the continuity of the scientific model established by Renan and Taine. In this respect, Renan's and Taine's thought can be defined as the crucible in which modern identities for the human and social sciences were forged around 1900.

Since then, their legacy has been largely eclipsed, while their approach to human subjects became outdated. To look back at their "untopical" science, therefore, implies to widen our prospects over the relatively recent past of the human and social sciences, to depart from a narrative written in terms of "disciplines," in order to consider other institutions, other scientific roles, and other forms of science. Such a study, as this essay hopes to demonstrate, could help us to identify some formative tensions which still characterize the modern academic fields of human and social sciences.

Obscured Legacy, Invisible Science

One of the reasons why Renan's, Taine's, and Maury's science of man has been forgotten lies in the strategies chosen by the founders of the modern human and social sciences around 1900. They insisted on the university as the sole institutional basis for the new disciplines, on profession, and on specialization. Even when they credited the intellectual legacy of the old generation, they criticized the former practices as amateurish dilettantism, blurring the lines between science and literature, between objective knowledge and ideology, etc.

A radical strategy of obliteration can be observed in Durkheim's early works, whereas historians adopted a more complex attitude. In *Les Maîtres de l'histoire*, for example, Gabriel Monod, the founder of the *Revue historique* (1876), a new periodical dedicated to the transformation of history into a professionalized and specialized academic field, accepted the intellectual and methodological legacy of Renan and Taine, but still advocated different practices and new institutions for history (Monod 1894).[2]

Émile Durkheim's early papers, dedicated to the genealogy and definition of the new sociology he intended to institutionalize in French universities, bear evidence of a more radical way of dealing with such a legacy. In 1900, retracing the development of sociology in France in the nineteenth century, Durkheim did not mention either Taine or Renan among its forerunners (Durkheim 1900). The dominant position Taine and Renan had granted to psychology in their science of man was indeed contradictory to Durkheim's definition of sociology as a science, which definition was warranted by the specific nature of the "social facts" it had to explain. According to this definition, Durkheim rejected what he called "subjective" sociology. By this epithet, he intended both to reject speculative philosophies of history (such as Comte's and Spencer's) and to bar definitions of social phenomena as caused by ideas and emotions which were the proper objects for social enquiry. Such a criticism was directed against John Stuart Mill, but it could also apply to Taine, who had commented upon and pursued Mill's logic of moral sciences (Durkheim 1975/1895; Taine 1864). Thus, if Durkheim accepted Taine's epistemological legacy as a fruitful philosophical synthesis between empiricism and rationalism (Durkheim 1897), he never recognized him as a direct inspiration for his sociology. On the contrary, Durkheim's social science was founded against Mill and Taine's psychological science of man.

Durkheim's position on Renan is similar. As a young professor, in one of his first essays he criticized Renan's political and social elitism (Durkheim 1967/1883).[3] But his early project of building a science of "moral facts" (a phrase coterminous with "social facts" in Durkheim's early works) could not be conceived without referring to Renan's philosophy and history of religion. Durkheim indeed implicitly referred to Renan's thought in 1887, in a review of *L'irréligion de l'avenir, Étude sociologique* by Jean-Marie Guyau. He criticized comparative history of religion (comparative mythology) for studying creeds and practices only as events taking place in individual consciousness, whereas they should be dealt with as social facts (Durkheim 1887, 299). Durkheim reiterated this critique, explicitly quoting Renan, in *The Elementary Forms of Religious Life* (Durkheim 1912, 102).

Durkheim's critical attitude, therefore, established the scientific agenda for French sociology against Renan's or Taine's science of man. It explains why neither belongs to the pantheon which the French sociological discipline has created for its own use during the twentieth century. Neither manuals nor academic syntheses dealing with the history of sociology mention them as important ancestors.[4] But their obscured legacy and the process itself of this revision of the past can be retraced in Durkheim's early texts. Strategically obscured at the beginning of the twentieth century, the legacy of Renan, Taine, and Maury is also clouded by the fact that their intellectual projects cannot be understood within the boundaries of the specialized fields, such as history, sociology, or psychology, which were subsequently founded as autonomous disciplines in modern universities. The historical, or rather memorial, traditions invented within these fields have often fragmented one person into many unrelated intellectual figures.

Alfred Maury, for example, is remembered by historians of religion and popular culture. His book on the history of magic is quoted by Michelet as one of his main sources in *Satanism and Witchcraft* (or Sorceress) (Maury 1860; Michelet 1966/1862, 308f). As a curator of the National Archives deposit, Maury is also well known to librarians. As an inspiration for Freud's dream analysis, he is remembered by historians of psychology and psychoanalysis. But, until very recently, no attempt had been made at understanding his work as a whole.[5] The situation is similar in regard to Taine's and Renan's intellectual descent. The former is known in the fields of literary criticism and æsthetics (Taine 1858, 1880); he is known to historians as the author of a conservative analysis of the French Revolution (Taine 1875/1893); his book *On Intelligence* is considered by psychologists as one of the founding treatises of French scientific psychology (Taine 1870).[6] But the relation between Taine's literary and art criticism, his history, and his psychology has not been studied. Renan is well known for his work on the history of religion[7]; he is also considered in literature, as the author of an autobiography, a classic in the French literary canon (Renan 1883); his studies on language are remembered in the field of linguistics (Renan 1858); his paper "What is a Nation?" is very often commented upon by specialists on nationalism. But the links between these different aspects of Renan's work have not been a subject for research.

Histories written in terms of disciplines, therefore, produce fragmentation. Did Maury, Renan, or Taine truly have multiple intellectual personalities? Did they see their writings as belonging to different but complementary fields? Or did they see them as belonging to a single global scientific field? A point of view oriented by the modern concept of discipline forbids answering such questions. Moreover, it prevents historians of science from understanding the formula used by these authors, which now sound strange because they relate scientific fields today considered as distinct. What was Renan's intention when he described his project for a science of mankind as a "historical psychology" (Renan 1995/1890, 223); what did Taine mean when he assessed that history was "fundamentally, a problem of psychology"[8] (Taine 1881–82/1864, xlv)? Fragmented by subsequent disciplinary traditions, such scientific agenda have become unreadable. Indeed Renan's, Taine's, and Maury's *science de l'homme*, as it became "untopical" within modern human and social sciences, also became invisible.

The "a-Disciplinary" Science of Renan, Taine, and Maury

The aim of this paper is precisely to look back at such a blind spot in history. By adopting an a-disciplinary point of view, it intends to analyze a synthetic cluster of fields which are now considered as distinct, and to restore Maury's, Taine's, and Renan's work as a coherent, nondisciplinary, intellectual project. Such a study reveals that, beyond their differences, these authors had many similarities.

Primarily, they explicitly depicted their work as a unified, single project. They named it either *science de l'humanité* (science of mankind), *science de l'homme* (science of man) in the singular or plural form (Renan and Maury), or psychology (Taine). Secondly, they defined this project as philosophical. They criticized ancient, abstract metaphysics and explicitly departed from it in order to reform philosophy on the new basis of observation and experimentation (Renan 1947–1961/1860, 681–714; Taine 1857). In this respect, they all referred to positivism as an inspiration: Renan (Petit 2005, 2003) and Maury (Bowman 1978; Richard 2007) commented on Auguste Comte's philosophical system; Taine derived his main inspiration from John Stuart Mill's *System of Logic* (Taine 1864). Their *science de l'homme* was therefore a philosophical venture, as it had been, at the end of the eighteenth century, for the *Idéologues*.

This statement highlights the well known fact that the origins of modern human and social sciences lies in philosophy. But it suggests that the secession of such fields from philosophy is quite recent, and that the demarcation lines between them is perhaps not as impassable as the segregated modern academic curricula would indicate. In France, for example, the discourse about autonomy, professionalization, and academic institutionalization of the new human and social sciences was often grounded in the rejection of philosophy. Around

1900, historians did compete against philosophers in order to reform curricula.

They claimed, with some success, that history was the most qualified science to train future citizens in a secularized, democratic state, and therefore asked for prominence in universities and in high schools (Fabiani 1988; Richard 2006). This founding conflict has resulted in the oblivion of philosophy as a recent root of historical discipline. A very similar process, with similar consequences, took place with Durkheimian sociology.[9] Since then, omissions in disciplinary memories have consolidated distinctiveness in academic curricula, but it has not prevented human and social sciences and philosophy from developing a reciprocally critical relationship.

Another characteristic of Maury's, Taine's, and Renan's approach to human subjects is that they gave prominence to psychology, believing that psychological factors were the main cause of human phenomena. I suggest naming this fundamental bias "psychologism,"[10] in which psychology was defined both as the basic level where complex phenomena could be deconstructed (analyzed) into more simple elements, and as a unifying method of explanation. This explains why Renan, Maury, and Taine used psychology as a demarcation criterion between human and natural sciences. Renan put the emphasis on a distinction between subject-matters: because of consciousness, or "spirit" or "soul," man was drastically different from natural objects. His science, wrote Renan, "did not deal with man considered as an animal" (Renan 1876, 1228). He distinguished two coeval worlds, "the physical world" and "the moral world," implying two separate orders of science, natural sciences, on the one hand, and sciences of mankind, on the other (Petit 1984; Renan 1947–1961/1860, 695f). As Maury and Taine adhered to the thesis of psycho-physiological parallelism, they believed that mental activities had a physiological basis. They could not, therefore, demarcate human from natural sciences through subject-matters. But they argued that these sciences differed from a methodological point of view: there was a "psychological method" which, though it was founded on observation and experimentation, was different from the naturalist approaches.[11] Such demarcations between human and natural sciences did not prevent analogies. But numerous comparisons did not imply identification; on the contrary, they often emphasized an essential distinction between these sciences. For Renan, Taine, or Maury, organicism was a metaphoric way of thinking, not a literal one (Richard 2004).

These considerations about the specificity of a science taking man as its object predate Wilhelm Dilthey's *Einleitung in die Geisteswissenschaften* (1883), and can be considered as a crucial step in the progressive definition of human sciences. To my knowledge, they display the first occurrence in France of the "great divide," warranting the modern identity of human and social sciences. Before them, such a divide had not been conceived by the promoters of a "science of man": neither the *Idéologues* nor the anthropologists following the Buffonian tradition of a natural history of mankind had grounded their scientific project in contrast with natural sciences.

Associated with the central position of psychology, a fourth characteristic of Renan's, Taine's, and Maury's science of man is that it is based on history. All of them were influenced by German philosophy, which was introduced in France both directly and through Victor Cousin's commentary, and they believed that to understand human phenomena, one had to study their genesis. Their research, therefore, always took a genealogical turn, and in many respects they can be considered as promoters of a French "historicism." Most of their writings, indeed, analyzed human phenomena in a historical manner. Alfred Maury studied the evolution of religion and popular superstition from classical antiquity to his days (Maury 1860); Renan's works on language or religion focused on the question of origins and early past (Renan 1858, 1863/1881); Taine's analyses of art, literature, and politics were historical (Taine 1881–1882/1864, 1875/1893).

In accordance with their creed in psychology as an ontological and methodological principle for studying human phenomena, Renan, Taine, and Maury carried out a synthesis between the now distinct fields of history and psychology. They borrowed tools from contemporary scientific psychology (which they also helped to establish), such as self-observation or observation of mental illness, and combined them with empathy in order to describe and explain the psychological motivations of actors in the past. In *Vie de Jésus*, Renan undertook to explain why and how Jesus' contemporaries believed in miracles. To describe their mentalities, he used explanations borrowed from contemporary psychiatry and from self-observations about the impact of landscapes on mind, realized during a trip in Palestine.[12] In *La Magie et l'astrologie*, Maury intended to explain why superstition, under the guise of new evangelist sects (such as Mormons), of belief in Marian apparitions or of creed in animal magnetism, for example, coexisted with modern rational science. In so doing, he used examples taken from his work, largely based on self-observation, on the psychology of dreams and mental disorders (Maury 1861; Richard 2007). Taine's *Origines de la France contemporaine* provided an explanation of the French Revolution in terms of passions and madness. The portrayals of the main Jacobin leaders were modeled on case studies taken from contemporary alienists, which Taine had analyzed in his psychology book *On Intelligence* (Richard 2008, 2002).

This manner of combining history and psychology in either a historical psychology or a psychological history has become outdated, and is, therefore, difficult to understand (and even to identify) by psychologists or historians trained within their now very separate fields. In order to re-establish the logic which was at the core of Renan's, Maury's, and Taine's science of man, one has to escape from modern disciplinary habitus.

Though it is not understandable from an anachronistic disciplinary point of view, and has become invisible, this intellectual project was coherent and exercised a strong seductive power on its contemporary public. It promoted a definition of man, or an anthropology, which had social, political, and religious implications. Phrased in a non-metaphysical, secularized, scientific, and psychological vocabulary, this definition showed human subjects as rational and irrational at the same time, and embedded in complex networks of causation which allowed for free will but also limited the power of reason (Burrow 2002). Such a definition seemed to echo the paradoxes of modern times and convinced Maury's, Taine's, and Renan's numerous readers. In the longer term, their definition of science became a fruitful basis for the modern foundation of human and social sciences. It allowed for crucial progress in methodology, and in the demarcation between human and natural sciences, ideas of which Wilhelm Dilthey was, among many others, well aware.

In relation to history of ideas, such a scientific project is, properly speaking, "a-disciplinary." It cannot be accurately described in the fashionable vocabulary of "transdisciplinarity," or in Ben-David's vocabulary, as the "hybridization" of different fields of knowledge. This would imply, in fact, that such separate fields existed before Taine's, Renan's, and Maury's time. Many examples taken from the previous generation of so-called "historians" or "philosophers," such as Jules Michelet, Edgar Quinet, or Victor Cousin, do not support such a statement. But if we take into consideration the social and institutional characteristics of the foundation of such a project, then

Ben-David's notions of "scientist's role" and "role-hybridization" become relevant.

Philosophers" and "Intellectuals": Taine, Renan, and Maury as Authors

Being by essence a-disciplinary, Renan's, Taine's, and Maury's *science de l'homme* was mainly developed outside academic institutions. Its institutional and social basis indeed differs from university departments and from organizations which now frame research and teaching in human and social sciences.

Before 1870, only Maury received institutional recognition and was appointed to a chair at the *Collège de France*, though he had no adequate degree in history or in philosophy. He had not completed his degree in medicine, and had studied law, dropping out before writing his doctoral thesis. He sustained himself by being appointed to various positions as a librarian. He began his scholarly career with his election to the *Académie des inscriptions et belles-lettres* in 1857. His reputation as a scholar, and his personal relationship with the Emperor, were sufficient for his nomination to the *Collège de France* in 1862. But part of his intellectual work, especially his research on dreams, was not broadcast through his lessons at the *Collège de France*.

Suspected to be opponents of the regime, Taine and Renan, in distinction, did not obtain any position at the *Collège de France* or at the Sorbonne, in the *Faculté des lettres*. Coming from a modest middle-class background, they aimed at an academic career, which they were refused. Taine was a brilliant student at the *École normale* when he failed the *agrégation de philosophie* because he was suspected by the spiritualist members of the jury of being a materialist. This exam would have allowed him to teach in the best high schools before completing a doctoral thesis and entering a *faculté des lettres*. He was, therefore, appointed to second rate schools, in small cities, far from Paris. Realizing that he would not succeed in the academic career he had aspired to, he left teaching in 1852 and never became a professor. At first, Ernest Renan was more successful. He was originally educated at the Saint-Sulpice Seminary before losing his sense of religious calling. He then began a program in philosophy and studied oriental languages. But he never intended to teach in high schools, aiming at a chair in philology at the *Collège de France*. He was appointed to this chair in 1862, but was dismissed after his first lesson. Already suspected of heterodoxy by the influential Catholic Church, he presented Jesus as a man. This lecture was considered a challenge to law and order, and Renan's class was immediately suspended. It resumed only in 1870, after the fall of the Second Empire. Neither Taine nor Renan was wealthy enough to live on private means. They therefore chose the best opportunity that was opened to well educated middle-class young men in mid-nineteenth-century Paris, a career as publicists and authors (Martin 1981).

Maury, Taine, and Renan were lucky enough to collaborate in influential periodicals, such as the *Journal des débats*, a moderate liberal daily newspaper, and the *Revue des deux mondes*, a review founded in 1829 on the model of the British *Edinburgh* and *Westminster Reviews*. Aimed at the educated bourgeois urban elites of their time, these periodicals established Maury's, Taine's, and Renan's intellectual reputation through articles and through reviews of their own main writings. At the same time, Renan's and Taine's ambition met the commercial interest of two dynamic publishers, Louis Hachette and Michel Lévy, who were transforming their businesses into capitalist editorial ventures (Mollier 1994, 1999) and were in quest of new best-selling authors. Renan's *Vie de Jésus* (1863) was explicitly written for a broad, non-specialized audience. It created a sensation in 1863 and a shortened, less expensive, popular version, *Jésus*, was written in 1864. The book became the biggest bestseller in nineteenth-century France; it carried Renan's ideas well beyond academic circles or educated male bourgeois readers, among middle-class, provincial readers, among women, and even among educated popular elites. It also made Renan's material fortune, and greatly increased Lévy's commercial profit. Taine began his collaboration with Hachette by writing a travel guide for a new collection which was sold in railway stations. But the *Voyage aux eaux des Pyrénées* (1855) turned out as a literary narrative, and met a large audience as such. Taine soon became known as an "impression" writer. The *Voyage aux Pyrénées* was reissued in 1858 in an amended version stressing its literary ambition, and in 1859, in a luxury edition with new illustrations by Gustave Doré. Taine became a popular socialite, regularly attending salons and literary circles such as the *dîners Magny* (Caramaschi 2001; Martin-Fugier 2007). He gave a satirical narrative of his experience among Parisian high society and salons in his *Notes sur Paris: Vie et opinions de Thomas Graindorge* (1867).

Such an institutional basis for the *science de l'homme* implied adaptation in both style of writing and the definition of the scientist's "role." For example, attending salons blurred the limit between "savant" and "socialite," and implied ability to write in a fashionable style. Taine became famous as a chronicler of high society. Maury's science of dreams gave birth to a fashionable social practice of writing down, telling, and discussing dreams among educated men. Renan's *Vie de Jésus* was the sensation of the 1863 season, and even the fashionable beauties of the "demi-monde" displayed it on coffee tables and pretended to read it. Selling books and writing articles for the general press also blurred the difference between the roles of scientist, journalist, and writer. Having no other mean of subsistence before 1870, Renan and Taine became authors.

Carefully marketing their image as martyrs of the Imperial regime, without going far enough to meet censorship or prison, they also broadcasted their work as opposing political, religious, and philosophical conservatism. Together with Maury and Littré, they promoted a positivist definition of man, opposed to spiritualism, weakening the distinction between science and politics. Such a scientific, æsthetic, social, and political role is not easily defined. The best word to describe it is probably "author," but this role also fits the later definition of the "intellectual" as organically integrated in his social, cultural, and political environment. Renan, Maury, and Taine could therefore be seen as intermediary, and largely forgotten, figures in the social and cultural genesis linking the eighteenth-century philosophers to the twentieth-century intellectuals, whether "integrated" (Gramsci) or "engaged" (Sartre).

Conclusion: Some Constitutive Tensions in the Human and Social Sciences

In France, at least, such a hybridization of styles and roles is the crucible from which the human and social sciences would later emerge. Renan, Maury, and Taine obtained recognition and prestige during the first decades of the French Third Republic. They died at the beginning of the 1890s, a few years before the 1896 Reform Bill which created the modern universities in France.

As the promoters of the new disciplines dealt with their legacy, they departed from their synthetic, a-disciplinary views, and from their philosophical prospects, in order to advocate specialization and professionalization by training. But at the beginning of the twentieth century, Taine's and Renan's role model was not yet completely outdated. Some social or human scientists, such as Gabriel Tarde, Gustave Le Bon, or Henri Berr, for example, still hoped for a synthetic science of man, still wrote in a polygraphic manner, and still addressed the general public. The competition between the academic, professional, and scientific role and the worldly, literary role of the publicist fuelled sharp controversies. In the 1910s, the Agathon affair, exemplifying the rivalry of conservative writers and academic (republican) scholars over higher education and over the production of knowledge about society, is, for instance, more easily understood when placed in such a context as well as in the framework of political conflicts.[13]

In the twentieth century, the academic role model gained prominence. Its grounding values became a *Leitmotiv*, to which referred, for example, Marc Bloch's plea for the "historian's craft" (Bloch 1997/1941). But the philosophical, anthropological, and synthetic tone of the *science de l'homme* never completely disappeared. As an ideal, it remained central to the ways in which some writers of the last century defined and established themselves as "intellectuals." This tension between the process of specialization in the universities, and an ideal synthesis which still invigorates some practitioners in the field of the human and social sciences, is important to understand as the context of the transformation and the success of the academic disciplines, but also as the background for the dead ends and the frustrations of both the scientists and their public. This argument is, in my opinion, a key to any fruitful debate on the future of universities. In this respect, looking back at the untopical, outdated, approach to human subjects by Taine, Maury, or Renan helps widen our present perspectives, especially when we also take into consideration the social and institutional characteristics of the foundation of such projects.

Based on the close analysis of some historical cases, and without denying the specificity of the past, this essay supports an argument about the present and hopes to encourage a critical examination of specialization in modern universities. The kind of scholarship fostered by the "role of the intellectual" seems to demonstrate the fruitfulness of approaches to human subjects which are not limited by disciplinary boundaries. It could help us re-evaluate, by comparison, the limits of disciplinary approaches which characterize most teaching and most research in today's universities.

It could also highlight the fact that a large part of the essays in philosophy, sociology, history, or psychology, which are read by the general public, are not produced by researchers working within academia and recognized as such by peers, and that the image we have of our fields and of our "craft" is not necessarily shared outside universities.

Notes

1. The modern university system was established mainly between 1870 and the Liard law in 1896. Cf. Joseph Ben-David (1977) and Christophe Charle (1994).
2. Cf. Nathalie Richard (2006).
3. See also Marcel Fournier (2007, 71).
4. For recent references, see Heilbron (2006) and Mucchielli (2006).
5. Such is the aim of Jacqueline Carroy and Nathalie Richard, Alfred Maury.
6. For a recent history of French psychology, cf. Carroy et al. (2007); see also Ian Dowbiggin (1990).
7. Ernest Renan (1863/1881), 7 vols., 1863–1881 (includes *The Life of Jesus* (1863) as its first volume).
8. My translation; in the following, this applies to all translated terms or quotes of French texts.
9. Mucchielli (2006) gives a narrative of this conflict. For more a complex story about the relationship between sociology and philosophy around 1900, see Brooks (1998). For the case of French psychology, cf. Jacqueline Carroy et al. (2006) and Carroy (2006b).
10. In this meaning, different from its pejorative and essentially epistemological use in twentieth-century philosophical debates, the neologism was applied to Renan around 1900; cf. Théodore Ruyssen (1908).
11. See, for example, the chapter entitled *Ma méthode d'observation* in *Le Sommeil et les rêves* by Alfred Maury.
12. Ernest Renan (1863), particularly chapter 16, *Miracles*; cf. Nathalie Richard (2003).
13. In 1911 and 1913, two young writers, Henri Massis and Alfred de Tarde (Gabriel Tarde's son), published, under the name Agathon, two pamphlets on *L'Ecrit de la nouvelle Sorbonne* and *Les Jeunes gens d'aujourd'hui*, in which they stigmatized the decline of classical culture in universities, the rise of "German" science, and therefore the decay of national tradition. Cf. Gisèle Sapiro (2004, 83–106) and Heilbron et al. (2004).

References

Ben-David, J. "Role and Innovations in Medicine." *American Journal of Sociology* 65, no. 6 (1960): 557–68.

______. 1971. *The Scientist's Role in Society: a Comparative Analysis.* Englewood Cliffs, NJ: Prentice Hall.

______. "Science and the University System." *International Review of Education* 18 (1972): 44–60

______. 1977. *Centers of Learning: Britain, France, Germany, United States.* McGraw-Hill.

Ben-David, J., and Collins, R. "Social Factors in the Origins of a New Science: The Case of Psychology." *American Journal of Sociology* 25 (1966): 828–843.

Bloch, M. 1941. *Apologie pour l'histoire, ou Métier d'historien.* Paris: A. Colin; English translation, 1984. *The Historian's Craft.* Manchester: Manchester UP.

Bowman, F. P. "Du romantisme au positivisme: Alfred Maury." *Romantisme* no. 21–22 (1978): 35–43.

Brooks, J. I. 1998. *The Eclectic Legacy. Academic Philosophy and the Human Sciences in Nineteenth-Century France*. Newark: University of Delaware Press.
Burrow, J. W. 2000. *The Crisis of Reason: European Thought, 1814–1914*. New Haven, CT: Yale University Press.
Caramaschi, E. 2001. "Au dîner Magny: Taine sous l'œil des frères Goncourt." In *Roman, critique, lecteur en France au XIXe siècle*, Firenze, Polystampa, 361–364.
Caro, E. 1864. *L'Idée de Dieu et ses nouveaux critiques*. Paris: Hachette.
Carroy, J. "Dreaming Scientists and Scientific Dreamers: Freud as a Reader of French Dream Literature." *Science in Context* 19, no. 1 (2006): 15–35.
______. 2006. "Philosophie et psychologie au XIXe siècle en France. Taine, Ribot et la *Revue philosophique*." In *Unité et globalité de l'homme. Des humanités aux sciences humaines*, M.-L. Pelus-Kaplan (dir.), 79–91. Paris: Syllepse.
Carroy, J., and Richard N., dir. 2007. *Alfred Maury, érudit et rêveur. Les Sciences de l'homme au milieu du 19e siècle*. Rennes: PUR.
Charle, C. 1994. *La République des universitaires*. Paris: Seuil.
Dowbiggin, I. "Alfred Maury and the Politics of the Unconscious on Nineteenth-Century France." *History of Psychiatry*, 1 (1990): 255–287.
Dupanloup, F. 1874. *Avertissement à la jeunesse et aux pères de famille sur les attaques dirigées contre la religion par quelques écrivains de nos jours, par M. l'évêque d'Orléans...* (1863), *in Nouvelles œuvres choisies*. Paris: Plon, tome 2, 23–180
Durkheim, E. "Discours aux lycéens de Sens (1883)." *Cahiers internationaux de sociologie* 43 (juillet-décembre 1967): 25–32; see also Fournier, M. 2007. *Émile Durkheim*. Paris: Fayard.
______. "De L'irréligion de l'avenir." *Revue philosophique* 23 (1887): 299.
______. "L'empirisme rationaliste de Taine et les sciences morales." *Revue blanche* 13, no. 101 (1897): 287–91.
______. "La sociologie en France au XIXe siècle." *Revue bleue*, 4e série XIII (1900): 609–613, 647–652.
______. 1912. *Les Formes élémentaires de la vie religieuse. Le système totémique en Australie*. Paris: Alcan; English translation, 1915. *The Elementary Forms of the Religious Life: A Study in Religious Sociology*. New York: Macmillan.
______. 1975. "L'état des études sociologiques en France (1895)." In *Textes 1. Éléments d'une théorie sociale*, 73–108. Paris: Éditions de Minuit.
Fabiani, J.-L. 1988. *Les Philosophes de la République*. Paris: Éditions de Minuit.
Heilbron, J. 1995. *The Rise of Social Theory*. Minneapolis, MN: University of Minnesota Press.
Martin, M. "Journalistes parisiens et notoriété (vers 1830–1870)." *Revue historique* 269 (1981): 31–74.
Martin-Fugier, A. "Convivialité masculine au XIXe siècle: les dîners Bixio et Magny." *Romantisme, Revue du XIXe siècle*, 137 (2007): 49–60.
Maury, A. 1860. *La Magie et l'astrologie dans l'antiquité et au moyen âge ou Etude sur les superstitions païennes qui se sont perpétuées jusqu'à nos jours*. Paris: Didier.
______. 1861. *Le Sommeil et les rêves*. Paris: Didier.
Michelet, J. 1966(1862). *La Sorcière*. Paris: Garnier-Flammarion.

Mollier, J. -Y. 1984. *Michel et Calmann Lévy, ou la naissance de l'édition moderne (1836–1891)*. Paris: Calmann-Lévy.
______. 1999. *Louis Hachette (1800–1864). Le fondateur d'un empire*. Paris: Fayard.
Monod, G. 1894. *Renan, Taine, Michelet. Les Maîtres de l'histoire*. Paris: C. Lévy.
Mucchielli, L. 1998. *La Découverte du social. Naissance de la sociologie en France, 1870–1914*. Paris: La Découverte.
Petit, A. "Renan et la classification des sciences." *Etudes renaniennes* 55 (1984): 4–14.
______. 2003. "Le prétendu positivisme d'Ernest Renan." *Revue d'histoire des sciences humaines* 8, *Les positivismes*, 73–101.
______. 2005. "Histoire et philosophie de l'histoire de Renan." In *Ernest Renan aujourd'hui. Études de Lettres*, Corbellari, A. (dir.), 3, 83–109.
Renan, E. 1858. *De l'origine du langage*. Paris: Lévy.
______. "La métaphysique et son avenir." *Œuvres complètes* I (1860): 695–96
______. 1863–1881. *Histoire des origines du christianisme*, 7 vol. Paris: Lévy [includes *The Life of Jesus* (1863) as its first volume]; English translation, 1864–. *The Origins of Christianity*. New York: Carleton.
______. "Des services rendus aux sciences historiques par la philology." *Œuvres complètes* I (1876): 1228.
______. 1883. *Souvenir d'enfance et de jeunesse*. Paris: Lévy; English translation, 1883. *Recollections of My Youth*. New York: Putman's Sons.
______. 1947–1961 (1860). "La Métaphysique et son avenir." *Œuvres completes*. Paris: Calmann-Lévy I, 681–714.
______. 1995 (1890). *L'Avenir de la science*, 223. Paris: GF – Flammarion; English translation, 1891. *The Future of Science*. Boston, MA: Roberts Brothers.
Richard, N. "Le 'merveilleux réel': Maury et l'histoire des aberrations de l'esprit humain." In Carroy, J., and Richard, N. *Alfred Maury*, 73–88.
cf. Richard, N. 2006. "L'Histoire en France (1876–1896): les fondements idéologiques d'une réussite institutionnelle." In *Unité et globalité de l'homme. Des humanités aux sciences humaines*, edited by Pelus-Kaplan, M. –L., 53–78. Paris: Syllepse.
_______. "L'histoire comme problème de psychologie. Taine et la 'psychologie du Jacobin.'" *Mil-neuf-cent* 20 (2002): 153–172.
______. "Analogies naturalistes: Taine et Renan." *Espaces Temps, Les cahiers* (2004): 76–90.
______. "Science et récit. Modèles médicaux et littéraires dans l'écriture de l'histoire en France au milieu du XIXe siècle." *Écrire l'histoire* 1 (2008): 127–138.
Ruyssen, T "Psychologisme et sociologisme. Revue de philosophie religieuse." *L'Année psychologique* 15, no 1 (1908): 357–72.
Sapiro, G. 2004. "Le savant et le littéraire. Les hommes de lettres contre la sociologie durkheimienne." In *Pour une histoire des sciences sociales. Hommage à Pierre Bourdieu*, J. Heilbron, R. Lenoir, and G. Sapiro (dir.). Paris: Fayard, 83–106.
Taine, H. 1857. *Les philosophes français du XIXe siècle*. Paris: Hachette.
______. 1858. *Essais de critique et d'histoire*. Paris: Hachette.

______. 1864. *Le Positivisme anglais. Étude sur Stuart Mill.* Paris: Germer Baillière.
______. 1870. *De l'intelligence.* Paris: Hachette; English translation, 1871, *On Intelligence.* New York: Holt & Williams.
______. 1875–1893. *Les Origines de la France contemporaine.* Paris: Hachette; English translation, 1876–1894. *The Origins of Contemporary France.* New York: Holt.
______. 1880. *Philosophie de l'art.* Paris: Hachette; English translation, 1883–1884. *Lectures on Art.* New York: Holt.
______. 1881–1882 (1864). *Histoire de la littérature anglaise,* 5e édition, 5 vol., vol. 1, xlv. Paris: Hachette; English translation, 1871. *History of English Literature.* New York: Holt & Williams.

9

American Universities and the Stagnation of Knowledge

Liah Greenfeld

One of Joseph Ben-David's central ideas—and the central idea of the original sociology of science which, together with Robert K. Merton, he helped to establish as a subdiscipline in sociology—was the idea that science was resistant to the point of absolute impermeability to broader social influences (Ben-David 1971; Merton 1973). Extrascientific social factors, Ben-David insisted, could often explain changes in the problematic, timing of the solution of one or another scientific problem, and its place, but could never explain the nature of the solution itself, which was always determined internally—and thus did not systematically impinge on the validity of science. This made science different from all other forms of knowledge and symbolic creativity, allowing it to develop cumulatively, building on and superseding past achievements, and thus made it progressive. Progress was impossible in art, literature, philosophy, or religion, all forms of knowledge and symbolic creativity other than science being socially permeable, often allowing social genesis to interfere with internal evolution, redefining standards, and, while making short spurts of cumulative creativity possible, preventing progress over any lengthy period of time (Greenfeld 1987a).

There has been consistent opposition to this view, within sociology of science as well as without, in critical literary theory, for instance, and especially in such areas of social criticism usually related to it as gender, Afro-American, and Latino/a studies, which became fashionable only by the very end of Ben-David's career (Barnes 1974; Bloor 1991; Lefkowitz and Rogers 1996; Lefkowitz 2007). Practitioners of these socially critical studies have viewed science as a particularistic perspective of the white males of European origin, and, seeking to validate the intellectually underrepresented particularistic perspectives of the minorities they represented, denied its claim to universal validity for ideological reasons. In addition, they resented it on personal grounds, for they obviously insisted that the knowledge about the intellectually underrepresented minorities they produced was itself not particularistic and to accept that science is the only nonparticularistic form of knowledge would mean being judged by standards of scholarship which they could not achieve (Lefkowitz 2007). (It is not surprising that the same objections to science were—in connection with genetics—raised in the Soviet Union, where science of the class enemy, the bourgeoisie whose European, if not white or male, character was often remarked on, was by definition bad science. But it is surely surprising to discover similar objections in the ideologically-opposed to contemporary social critics Germany, where too—in this case, in regard to physics—racially alien science was suspect [Forman 1971; Joravsky 1970].)

The opposition to Ben-David's position within sociology of science likely has been motivated by considerations of an entirely different nature. In a certain sense, sociologists are believers in the universality of science themselves. The characteristic of science, they believe, is that its propositions apply universally across all the phenomena belonging to the same category. The difficulty for them was the argument that science alone of all areas of knowledge or symbolic creativity was impenetrable to social influences; it looked to them suspiciously unscientific, as if playing favorites, privileging science for ideological reasons.

The argument, however, is sociological, not ideological. The reason for the extraordinary impermeability of science to social influences, paradoxically, is a function of its social structure or its structure as an institution, whose specificity was captured by Ben-David (1971) in his concept of the "scientist's role." Like every social role, the "scientist's role" is the pattern of individual behavior prescribed by a set of social norms and oriented toward a specific socially approved goal, that is, a goal set by the society as a value. In science, this goal is the satisfaction of the individual curiosity about the empirical reality. Science became an institution when this natural human propensity (which, obviously, has always existed) was set by a society as a social value, the social validation making it a career opportunity or, as Ben-David would say, "a magnet for talent." This momentous social development occurred in England in the course of the seventeenth century and it changed dramatically the development of scientific knowledge, making possible its progress (Greenfeld 1987b; Merton 1938).

If before, curiosity about the world, as it was, was likely to get one into trouble and science was as dangerous a pursuit as witchcraft (think of Giordano Bruno, who was burnt at the stake, or even Galileo, who was only silenced and spiritually broken), now curiosity was lauded as a sign of mental superiority and as socially useful, and science became one of the high roads to status. The social validation of empirical curiosity logically led to making normative a certain set of behaviors which were conducive to its satisfaction. It is this normative structure that we recognize by the name of "the scientific method" and which consists, in the words of Karl Popper (1963), of "conjectures and refutations," logical formulation of informed hypotheses and their methodical, systematic, testing against empirical evidence. This method (this set of prescribed behaviors) sensitizes the scientist to inconsistencies and produces the mechanism of scientific change—their progressive elimination. Every succeeding hypothesis or theory in a field derives from the one preceding it and the inconsistencies (or, "anomalies" [Kuhn 1962]) discovered between it and the evidence with which it was confronted, resolves these inconsistencies, and thus represents a greater approximation to truth (Popper [1963] calling this "a lesser falsity content" and, therefore, "a greater truth content"). It is because of this peculiar structure of scientific activity as an institution—the special value to which it is oriented, the special normative structure required by this value, the sociopsychological mechanism of change built into this normative structure—not because of any inherent merits of scientific knowledge, that science follows the law of its internal evolution, virtually unswerved by the intrusions of social genesis, and it is this institutional structure that, therefore, justifies its claim to universal validity.

The original sociology of science which Ben-David helped to establish and represented is the only explanation for the most striking empirical characteristic of science—its progressive development in the course of the last four centuries (the time since its institutionalization) and the sustained accumulation of reliable objective knowledge. The explanatory power of this hypothesis, which decades of opposition failed to refute, has naturally encouraged further hypotheses regarding structural determinants of scientific productivity, pertaining not only to the abstract structure of science as a social institution, but to concrete structures of specific organizations in which science has existed in different societies, including, first and foremost, the university. Ben-David (1977) was the first to suggest such extension of his central idea in his analysis of styles and levels of scientific productivity in Britain, France, Germany, and the United States in his second very important book, *Centers of Learning*. The structure of American research university, with its graduate departments, he argued there, was particularly conducive to the advancement of scientific knowledge, which explained the leadership of the United States, though a latecomer to the race, in virtually every field of scientific endeavor in the twentieth century—after this characteristic type of institution emerged.

It was true, indeed, that the United States has been an undisputed center of the scientific universe, when *Centers of Learning* was written, already for some half a century or so, in many areas replacing Germany which made it impossible for its best minds, Jewish and therefore racially unacceptable, to contribute to the advancement of knowledge at home. It was true, too, that this American ascendancy coincided with the spread of the research university model throughout the vast country. But it was not the entire story. Quite apart from the possibility (considered by Ben-David who was a very careful scholar, unwilling to sacrifice evidence even to a most winsome argument) that the shift of the scientific center to the New World was as much a result of the absorption of the Jewish scientists fleeing from Germany and Eastern Europe (in particular, physicists and mathematicians) in previously mediocre departments of American universities (see, for instance, Nasar 1998), or of the devastation of Europe by World War II, from which the United States emerged with a stronger economy (in absolute as well as in relative terms) than it had before the conflict, and therefore able to lavish funds on its scientists, as of the nurturing structure of the research university, a very important factor was left out of account.

I do not want to suggest by any means that my dear and only teacher missed what was plainly visible. This factor, though extremely important, was difficult to perceive at the time, perhaps, from Ben-David's point of view, even impossible. Ben-David was a sociologist, he belonged to the first generation of Israeli social scientists—a small group of talented, dedicated people, most of them European immigrants and often (like Ben-David) members of families who perished in the Holocaust. Their choice of their unlikely for the young struggling country profession was a reflection of their patriotism and the belief that objective knowledge of social reality was as important as that of physical and biological realities. They did not doubt that such knowledge was possible and that social sciences were as much sciences as physics or biology. Therefore, they turned a blind eye to the fact that *social sciences,* without a single exception, did not display the most striking characteristic of science—the unusual pattern of development that distinguished it from all other intellectual activities—and *did not progress.* And this stagnation of the social sciences was the chief empirical indication that a central factor was missing out of the account of scientific progress.

Social sciences, from the very day of their institutionalization as such, that is, from the moment they were declared sciences, existed in the framework of research universities (Greenfeld 2001, 428–72; Greenfeld 2002). If the structure of research universities itself—on top of the social approbation of the goal of science, implicit in its institutionalization—promoted the development of science, social sciences should have been augmenting objective knowledge about social reality by leaps and bounds. No such thing happened, however. Social scientists have certainly produced a very large number of publications and, probably, have a greater amount of printed pages to their credit than the other sciences. In this sense, they have been very productive (or destructive of trees). But their product for the most part has not been objective knowledge about social reality. They could not produce such knowledge because work in them would only occasionally represent the application of the scientific method of conjectures and refutations to this reality, or logical formulation of hypotheses methodically tested against relevant empirical evidence. Much more often it consisted, instead, of an exercise, that could only be characterized as scholastic, of constructing speculative models of what aspects of reality could have been were they based on certain principles (such as economic rationality, for instance) which did not obtain in fact, and then discussion, analysis, and, where the initial simplicity of the principles allowed this, mathematical development of these models. These models were never intentionally tested against empirical evidence or actual examination of the relevant empirical reality, as prescribed by the scientific method. Wherever circumstances submitted them to such test (as has repeatedly happened in the most influential social science—economics), these models were proven false. They would not be then rejected, however, but only put aside and revived time and again after the unfortunate circumstances, and with them empirical reality, faded from memory. In economics, this has been the characteristic pattern, but political science and sociology, too, have often followed it.

When they have not, these disciplines have divided their attention between collection of quantitative information on fashionable topics likely to be funded (such as patterns of smoking among schizophrenics—a typical choice in the sociology of mental health, or patterns of voting in "transitional" societies—a fashionable preoccupation among ex-Sovietologists among political scientists) or open-ended explorations of specific cases or examples of contemporary social life. These kinds of studies, especially the latter, were often valuable, containing much previously unrecorded and maybe even unnoticed information that could be eventually used in the testing of scientific hypotheses—as in fact happened to the information collected during the "natural history" phase of biology. The actual contributions of the social scientists trained as such in the graduate departments of American research universities and elsewhere to the common stores of knowledge were almost exclusively of this "natural history" nature. Collection of information, however, is not science: no amount of information can increase "the content of truth" in our knowledge, which accumulates by deduction and not by induction. Moreover, the "natural history" phase of the human (obviously including social) sciences did not begin with the institutionalization of the social sciences but some 2500 years ago, and, despite the accumulation of prodigious amounts of information, led us no further in our understanding of the reality to which all this information pertains than we were when we first started thinking of it logically. Only in the rarest of cases has the collection of information by the social scientists resulted in a theoretical advance, that is, in the articulation of an explanatory principle that could be applied beyond the particular case about which information was collected. Such was the case of Ben-David and his idea of "role-hybridization" in science, among other things. But science is a collective enterprise, to move forward it requires that such ideas be built upon, and Ben-David's ideas, like those of other exceptional individuals, were not even seriously entertained. (The idea of "role-hybridization" developed the earlier ideas of "social role" and "institution," ultimately derived from Durkheim and Weber, neither of whom were social scientists by training.) As a result, we find Ben-David and people like him in the position of natural scientists before the institutionalization of science: of course, they existed, but science did not advance.

You may have noticed that I am not stressing the pervasive politicization of the social sciences to which their failure to progress is occasionally attributed. Though absolutely rejected by the norms of science (and incompatible with the scientific method), it has indeed accompanied the social sciences throughout their triumphal march through the institutions of higher learning, making their scientific credentials even more dubious. This politicization was so rampant in the 1970s when Ben-David worked on the *Centers of Learning*, that he became concerned for the future of American universities. Nevertheless, it cannot be considered a countercurrent to the structural influences of the latter, presumed to be beneficial to the scientific progress, but is, rather, an expression of the tendency toward specialization which the structure of the research university favors. More important in making the advancement of knowledge in the social sciences unlikely may have been the frankly careerist considerations of the first generation of social scientists in the United States, which figured in the motives for the foundation of the academic disciplines of economics, political science, and sociology at least as prominently as the interest in particular aspects of social reality (Greenfeld 2002).

In the American society, rapidly industrializing in the wake of the Civil War and as rapidly changing the traditional (pre–Civil War) class structure, these first social scientists—children of the traditional, displaced elite, educated in the faculties of philosophy (including everything that was not law, medicine, or theology) in traditional German universities—felt unappreciated and were centrally preoccupied with status. They were not curious about social reality, but dissatisfied with the society in which they lived and were attracted to science not by natural inclination but because in their time science more than any other pursuit connoted authority. When the great self-made men of wealth—people whom the first social scientists called "robber-barons" and heartily detested—began to fund and found research universities, these offended children of the gentry were the first to jump on the band-wagon and ensconce themselves in them. Because the closest they came to a study of social reality in Germany was history, history was the first "social science" in America. Because the new elite whose rise irked them the most was economic, the history on which they first focused was mostly economic history with some helping of social criticism. Because of this and also because it was possible to formulate certain philosophical propositions related to economic matters mathematically, which made them stylistically similar to propositions in physics—the crowned queen of sciences—economics was the first social science to branch off from history and establish itself in the research university alongside it. The fact that this momentous event took place during the first annual meetings of the American Historical Association—a group of first American historians splitting off this newly formed trade union to form the American Economic Association and declaring themselves first American economists—could easily be taken for fiction; it appears so neat a metaphor for the opportunism of these entrants into one of the most disinterested professions as to be almost incredible. Yet this is how it happened. And the birth of political science and sociology in very similar circumstances—American Political Science Association emerging during annual meetings of the American Economic Association, with some of the first American economists, ex-historians, becoming first American political scientists, and American Sociological Society seeing the world some years later during the meetings of the APSA—does not make it easier to believe. In a final proof that on some occasions one should not bother to write satire, life being the best satirist itself, this side-splitting development concluded with the establishment of the ASS, the giant brains among its founders only with some delay appreciating the joke they played on themselves.

The rapid differentiation of the newly conceived "social science" into separate disciplines, clearly, did not follow the "role-hybridization" pattern, which Ben-David found in biology and medicine (Ben-David 1960; Ben-David and Collins 1966), and was inspired neither by "anomalies" within previously accepted theories nor by the unorthodox training and special perspectives brought by individual researchers to an established area of research. It was not a series of stages in the internal evolution of science, in other words, but cases of unobstructed social genesis of forms of knowledge, which could not occur in science by definition. What allowed them to happen was the structure of the American research university with its graduate (and related undergraduate) departments for the training, or reproduction, of specialists in the sciences. The result was that an activity that was not science was institutionalized as science in the United States (and eventually in the rest of the world that followed the American model) and accorded the freedom and authority of science. This nonscience under the name of science, quite obviously, could not be progressive, and its institutionalization within the university was to have a woeful effect on college education with which the latter in America was entrusted (see Peter Wood's essay in this volume).

But the main reason behind the stagnation of the social sciences had nothing to do with the structure of research university or even that of science as an institution, which brings us back to the very important factor in the explanation of scientific progress, that was left out of the account. The thing is that even with empirical curiosity validated and scientific activity socially approved, rewarded, and optimally organized (whatever such optimal organization may be for the study of different aspects of reality—physical, biological, and that which was misnamed "social"), science cannot progress unless the philosophical, specifically metaphysical or ontological—pertaining to the nature of reality, system of thought on which it is necessarily based allows it to do so. The favorable philosophical framework is the fundamental factor in the development of science: whereas the institutionalization of science ensures the continuity of scientific activity over time, our beliefs about the nature of the world around us, on which all of our activities depend, divide our lived experiences into real and epiphenomenal, empirically accessible and empirically inaccessible, objective and subjective, material and spiritual. Insofar as science is concerned, all these dichotomies translate into legitimate subjects for scientific study and subjects that are not legitimate.

Since time immemorial (or for the historians among us, since fifth century BCE) reality has been conceived of in terms of such dichotomies, though the chief of them, that of material vs. spiritual realities or substances, did not usually correspond to the others. This correspondence, in fact, is a product of the institutionalization of science and, three centuries later, its elevation to the position of ultimate epistemic authority (which, among other things, presumed that science, necessarily based on our image of reality, should be the foundation of this image). This ascendancy equated the material with the real, objective, and empirically accessible, and, as a result, defined whatever is not material as empirically inaccessible, subjective, and unreal. Logically, this created a jumble:

> our experience is inescapably subjective;
>
> "empirical" is what we know by experience, but, obviously, most of what we know by experience is not material (make a little experiment and recollect all your experiences since this morning); what we know is the only reality for us, but much of what we know we do

> not know by experience, rather, we know it indirectly or based on someone else's authority or belief;
>
> defining empirical as material we have excluded most of our experience from scientific study and relegated it to unreality;
>
> conclusion: we concentrated our scientific efforts on what we cannot know by experience because we believe that only the material is real.

Such is the power of philosophy which we have thought we subjugated to science: it limited what science may study. Fortunately, in the process, we have forgotten what "material" in fact is and thus were able to include in the province of science a large sphere which could not be so defined. Unfortunately, we could not know this sphere by experience either.

It is not coincidental—though, apparently, it was assumed to be—that physics was the first science to advance rapidly when scientific activity was institutionalized and that it was the only one to make progress during at least the first two and a half centuries of modern science's existence. Physics developed because Western philosophy—the dualist ontological (matter/spirit) vision of reality—allowed for the reality of matter. When the development of modern physics began, as we all know, the reality of the spiritual was yet undeniable: it is obvious that Newton and other great scientists of the period would have assigned it causal priority. The definition of reality as material was to come later and precisely because physics was for so long the only science. But the science of today—biology—it should be remembered, did not begin to develop until after the publication of Darwin's *The Origins of the Species*, though the interest in biological matters was great and though the social approbation of curiosity extended to the curiosity of biologists as well. Biology failed to develop because the belief that reality consisted of two elements, the material and the spiritual (whatever was the relationship between them) forced upon it the necessity to envision life in these terms, and life fitted neither.

Life cannot, strictly speaking, be considered a material phenomenon. Matter is extended substance, it takes place in space. All the physical forces are spatial; the other fundamental category of thought, time, as we know, has no physical qualities and exists in physics solely as a measurement of spatial relationships. Life, in distinction, is a process, and as such it occurs in time. True, to every stage, or event, in the life process correspond states of matter which take place in space, but time has in biology a significance very different from physics: material states in the process of life utterly depend on *when* in this process they occur and cannot be understood at all in terms of their spatial relationships alone. Organic matter is matter dependent on time, matter which is, in other words, deprived of its autonomy and ruled by other than material laws. The dual vision of reality allowed for such possibility only if the laws governing matter in life were spiritual, and so biology stagnated between *mechanism*, which attempted to reduce it to physics, and *vitalism*, which equally vainly tried to infuse it with *elan vital* on which no one could get a handle.

Darwin's theory of evolution through natural selection transcended both these (materialistic and idealistic) possibilities. Implicit in it was a very different view of reality, as consisting of logically consistent but autonomous layers, life on top of matter and dependent on it only in that matter provided boundary conditions, outside which it could not exist and using material elements in its construction, but causally independent from matter and irreducible to these elements which in its use it transformed. Life, in other words, is an emergent reality (in the original, strong, sense of the concept of "emergence")—this conclusion, perceived by several thinkers already in the nineteenth century, was masterfully presented by Michael Polanyi (1968). The theory of evolution through natural selection implied it because it postulated a general biological law, referring to a process (by definition dependent on time), which was completely unrelated to any physical law but perfectly consistent with all physical laws. With this law, Darwin freed biology from physics without in so doing involving any mysterious transcendental forces and, in fact, announced an element of reality which was neither material nor spiritual: the organic element.

Among the very first thinkers who perceived the vast new horizons that this implicit revision of age-old philosophy has opened was Emile Durkheim , still one of the two greatest social scientists, whose work has not as yet been superseded. Unfortunately, very few understood him and saw what he saw. After Darwin, biology began to advance apace, but few biologists cared about the philosophical underpinnings of their discipline or were aware of how revolutionary was the image of reality which made their explorations conceivable. Biologists work, as most people do, peaceably in the framework created for them, taking it for the only one possible. Among philosophers of science, physics has remained the model, and materialism (one of the monist varieties of the traditional dual view of reality) was not questioned;

there were some discussions of emergence since 1920, but the general implications of Darwin's theory were never spelled out and, judging by contemporary interpretations (Dennett 1995), it is doubtful whether they have been ever appreciated.

The chief of these implications—that reality may not be what it was believed in the Western philosophical tradition: matter and spirit somehow related to each other, that it may, instead be composed of logically consistent but autonomous layers—and the realization that life is, in fact, best imagined as such a layer, a reality emergent out of matter but irreducible to it, could have been of vast significance to the social sciences. Had the social scientists in the early twentieth century heeded Durkheim's suggestion that, perhaps, humanity also is best considered such an autonomous layer, emergent out of animal life but irreducible to it, social sciences too might have had a century of rapid progress behind them. Of course, they would not have looked in such a case the way they look now: they would not be split into virtually unconnected disciplines of history, economics, political science, sociology, etc. (analogous to natural history and separate sciences of limbs, stomachs, and hearts, for instance, in biology), and maybe even would not be called "social" sciences (since society is not what distinguishes humanity from the rest of life). Instead we would, probably, have human sciences, focusing on the only truly empirical reality we know—our subjective experiences or *the mind*, which would, in turn, make them focus on *culture*. But, as it was, the first social scientists took physics for their model, the metaphysics which enabled physics to develop for their philosophy, and, knowing no biology, chose to focus on the eminently biological subject, the corollary of life, society. It is not surprising that absolute stagnation has been the result. And stagnation is not the worst of it.

The worst—the perversion of academic freedom in the social sciences—was actually contributed to this stagnant intellectual pool by the structure of the American research university and specifically by the institution of the academic department. Departments are structural counterparts of intellectual specialization. They acknowledge, certify, perpetuate, and encourage it. Now in some fields—in technical fields mostly, such as engineering, in which specialization amounts to the honing of a particular skill (and a skill may only be technical), specialization may be a beneficial and a necessary development. But in other fields it is not. Think, for instance, of medicine: it would be awful to find oneself in the hands of a cardiologist who is ignorant of/and would not take into consideration problems that are neurological, pulmonological, hematological, even psychological; a psychiatrist who would be ignorant of cardiology, etc. For this very good reason, physicians are exposed to all of medicine and begin specializing only after having acquired a thorough knowledge of the human organism as a whole.

In the spheres that have to do with the tremendously complex (and for reasons mentioned earlier still very insufficiently understood) subject of humanity, specialization—that is, the splitting of the field into separate disciplines isolated in departmental administrative units, which control trainees' certification and, as a result, have a tremendous influence on their careers—is particularly pernicious. Departments, by definition guardians of disciplinary dogmas, necessarily come to control scholarship of its individual members and exact heavy toll from those who dare to break lines. Such institutional control of thought makes intellectually controversial positions—thus intellectual innovation—personally dangerous and virtually impossible. Creativity dies. But since competition for status must go on, it takes the form of politicization in the sense of the prevailing norms of political correctness—zeal in pursuit or espousal of political dogmas becomes the preferred way of distinguishing oneself and building a career. This combination of absolute intellectual conformity with political conformity explains why in American social sciences, in particular, given the standards of political correctness in the United States since the 1960s, diversity of physical type and sexual orientation has been substituted for the diversity of ideas.

This disastrous effect of the American research university on scholarship in the social sciences is, unfortunately, equaled by its effect on the education of young Americans (see Wood, this volume). As Ben-David stressed, the American university has been from the outset entrusted not only with the training of future generations of scientists but with the general education of future generations of citizens. The college which was to provide such education, upon the inclusion within the research university, was also departmentalized with most departments consisting of the graduate, training, compartment, and a usually much larger undergraduate, educational, one, with the faculty hired by discipline to serve in both. It is evident that departmentalized college has not fulfilled this very important function and that it could not do so. American youth is dependent on the college for its general education. Given the state of the primary and secondary schools in the country, most of the children leave them ignorant, at best with little tidbits of learning. To subject them to a specialized college education is to close and impoverish their minds forever—and this means to impoverish and deprive of creativity the American culture in general.

One must distinguish between education and training. Training (which is what we do, among other trainees, to dogs) equips one with skills, namely, abilities to perform certain procedures. Thus one is prepared for particular occupations, the ways of competently earning one's living, whether it is that of a brain surgeon or blind-guide dog, a violinist or a circus animal. Obviously, the performance would be better, in the case of human as well as canine trainees, if the training develops a native ability, but we require that every engineer, doctor, or orchestra player be a genius no more than we do a special talent from our work-dogs. The difference is that we expect a brain surgeon and a violinist to be not only well-trained, but also educated. In distinction to training, education prepares a person for life by developing in him or her general orientations to enable them to act on one's own in a variety of circumstances and situations and irrespective of what one's particular occupation or role in the always limited economic sphere is. A dog is prepared for life genetically and carries such orientations in his blood. In traditional societies, in which birth equaled destiny, and conduct within every social position was prescribed by custom and religion, a child acquired them by watching others in one's immediate environment and at church. Only societies which offer one some choice as to self-definition have a need in formal education. The purpose of such education is, in the first place, to guide one in making this choice, that is, to help one to understand oneself.

It is to faculties of Arts and Sciences that students come in quest of education: to provide it is not the responsibility of professional schools (including the schools of education) which, instead, train students and equip them with skills for particular occupations. The quantitative burden of the educational task in the faculties of Arts and Sciences must necessarily fall on the social sciences and humanities, since natural sciences represent only a minority of the disciplines in these faculties and have a minority of students in their charge. Whatever one can say about humanities, the social sciences are completely unequipped to answer this demand. With the woefully limited education of the professors themselves, punished for thinking outside of their disciplinary boxes and taught to confuse academic freedom with the right of enforcing standards of the political correctness of the day, the best they can do is to clone skillful economists, political scientists, and sociologists. This is as good (or as bad) as raising skillful poets, musicians, or mathematicians, for none of whom "skillful" would imply a compliment, for it suggests no creativity, but only technical competence—with that difference that technical competence in the social sciences connotes exclusively the ability to talk the disciplinary talk and walk the disciplinary walk. The very large and increasing number of disoriented, unhappy undergraduates on our campuses, barely surviving their college experience with the help of prescribed antidepressants, illegal drugs, and alcohol, proves that such teaching them the tricks of the various trades does not do the trick.

Associated as they are now with entrenched vested interests, social sciences cannot be reformed. Nothing but a complete overhaul of the departmental system would do to turn social sciences into sciences in fact and/or into actual agents of education. The university is too formidable an institution to allow such an overhaul. Neither can one hope that the two tasks of actually advancing knowledge about humanity and actually educating young people in need of education might be accomplished outside the university, with the rise of new institutions, for, given the university's hold on certification and possibilities for the employment of the professoriate, no such institutions would be able to compete with it. The only alternative is to create transdisciplinary research centers and teaching programs within the university alongside the departments. Such centers must not simply unite faculty trained in different disciplines administratively, which would already protect them from the necessity to conform to disciplinary dogmas and ideological pressures, but focus on the common subject of all social sciences—humanity, transcend these disciplines as if no division between them existed, and begin as it were conceiving the science of humanity anew. They also must consist of senior and distinguished faculty, who have already proven themselves capable of such transcendence and who would not risk their careers to engage in such unorthodox work. The now fashionable interdisciplinary teaching programs in which professors trained in separate disciplines and working within the prescribed boundaries of one or another of them, mostly junior and naturally fearful of trespassing, teach students a *melange* of courses from different disciplinary perspectives will not do; the only thing such a smorgasbord of social sciences (or social sciences and humanities) can achieve is to acquaint students with their limited choices for graduate training, of which, from the intellectual and psychological (educational) points of view, each is precisely worth any of the others. To replace such interdisciplinary programs with trans-disciplinary ones or to establish the latter alongside the former, university administrators will have to be made to understand the impossibility of advancing the knowledge of humanity or of helping young people know themselves within the traditional structures they now administer. And this, of course, is a colossal undertaking in itself.

References

Barnes, B. 1974. *Scientific Knowledge and Sociological Theory*. London: Routledge.

Ben-David, J. 1971. *The Scientist's Role in Society*. Englewood Cliffs, NJ: Prentice-Hall.

Ben-David, J., and R. Collins. "Social Factors in the Origin of a New Science: The Case of Psychology." *American Sociological Review* 31 (August 1966): 451–65.

Bloor, D. 1991/1976. *Scientific Knowledge and Sociological Theory*, 2nd edn. Chicago, IL: The University of Chicago Press.

Dennett, D. C. 1995. *Darwin's Dangerous Idea: Evolution and the Meanings of Life*. New York: Simon & Schuster.

Forman, P. "Weimar Culture, Causality, and Quantum Theory, 1918–1927: Adaptation by German Physicists and Mathematicians to a Hostile Environment." *Historical Studies in the Physical Sciences* 3 (1971): 1–115.

Greenfeld, L. "Russian Formalist Sociology of Literature: A Sociologist's Perspective." *Slavic Review* 46, no. 1 (1987a): 38–54.

______. "Science and National Greatness in Seventeenth-Century England." *Minerva* 25, no. 1-2 (1987b): 107–122.

______. 2001. *The Spirit of Capitalism: Nationalism and Economic Growth*. Cambridge, MA: Harvard University Press.

______. 2002. "How Economics Became a Science: A Surprising Career of a Model Discipline." In *Disciplinarity at the Fin de Siècle*, chap. 3, edited by A. Anderson and J. Valente, 87–125. Princeton, NJ: Princeton University Press.

Joravsky, D. 1970. *The Lysenko Affair*. Cambridge, MA: Harvard University Press.

Kuhn, T. S. 1962. *The Structure of Scientific Revolutions*. Chicago, IL: The University of Chicago Press.

Lefkowitz, M. 2007. *History Lesson: A Race Odyssey*. New Haven, CT: Yale University Press.

Lefkowitz, M. R., and G. M. Rogers, ed. 1996. *Black Athena Revisited*. Chapel Hill, NC: The University of North Carolina Press.

Merton, R. K. "Science, Technology and Society in Seventeenth Century England." *Osiris* IV (1938): 360–632.

______. 1973. *The Sociology of Science: Theoretical and Empirical Investigations*. Chicago, IL: The University of Chicago Press.

Nasar, S. 1998. *A Beautiful Mind: The Life of Mathematical Genius and Nobel Laureate John Nash*. New York: Simon & Schuster.

Polanyi, M. "Life's Irreducible Structure: Live Mechanisms and Information in DNA Are Boundary Conditions with a Sequence of Boundaries above Them." *Science* 160 (June 1968): 1308–12.

Popper, K. R. 1962. *Conjectures and Refutations: The Growth of Scientific Knowledge*. New York: Basic Books.

10

The Sustainability Movement in the American University

Peter Wood

Higher education in the United States is today a troubled institution. One index of its troubled nature is the rise of an illiberal ideology that calls itself *sustainability*. I propose to characterize this movement and its underlying sociology in a manner that pays some debt to Joseph Ben-David's account of the constituent elements of the university. But sustainability is dangerous quarry and rather than pounce on it outright, I intend to sneak up on it circuitously.

Cerberus

A hunter, however, must know what he is looking for, so let me begin with a sketch of the beast. To start, sustainability, like the mythical dog Cerberus, has three heads, a fierce and watchful attitude, and stations itself on the boundary between life and death. Sustainability's three heads are environmentalism, green economics, and social justice. Its fierceness is directed toward those it suspects would trespass against its edicts on wastefulness and privilege. And it imagines itself perusing the boundary between Edenic renewal and an eco-apocalypse of unchecked global warming and widespread extinctions of plants and animals, including mankind.

The sustainability movement is, of course, not limited to the United States, but is rather robustly international. One of its founding moments was the report issued by the World Commission on Environment and Development, otherwise known as the Brundtland Commission, in 1987. That report, *Our Common Future*, has become one of its sacred texts.[1] It called for "sustainable development," which it defined as "development that meets the needs of the present without compromising the ability of the future generations to meet their needs." The Brundtland Commission was far from the first gesture in this direction. The UN's official list of global environmental agreements commences with the 1971 *Convention on Wetlands of International Importance especially as Waterfowl Habitat* which emerged from a meeting in Rasmar, Iran. Since then there have been over fifty such international agreements, including the *Rio Declaration on Environment and Development* agreed to at the 1992 UN Earth Summit; the establishment in 1993 of the UN Commission of Sustainable Development; the 1994 *Declaration of Barbados* on sustainable development of small island states; the *Kyoto Protocol* adopted in December 1997; the *Earth Charter* presented at the Hague in June 2000; the 1992 *Johannesburg Declaration on Sustainable Development*; and the commencement in January 2005 of the UN *Decade of Education for Sustainable Development*. Next up is the United Nations Climate Change Conference in Copenhagen, in December 2009.

The numerous reports, declarations, protocols, and agreements also firmly place "sustainability" in the realm of policy and politics. To the extent that Cerberus frequents the precincts of the university, he may be straying beyond the central precinct of the river Styx. The university's embrace of sustainability comes well after the ideology was elaborated by governments in international conclaves. It also comes many decades after the rise of environmentalism in its twin forms of popular mass movement and radical cause. In the United States, the mass movement aspect of environmentalism crystallized on April 22, 1970, the first Earth Day, which nationwide drew about 20 million participants to rallies against despoliation of the environment. The event came about as the initiative of U.S. senator Gaylord Nelson, and drew together environmental activists, college students, celebrities, and paid advertising agencies. It was meant to be international, and the organizers claimed to have gained the participation of 200 million people in 141 countries. It seems doubtful that all 200 million had a shared conception of what they were promoting, or that there was even much consensus among the 20 million in the United States, but mass mobilizations mean something, even if it is no more than an unmistakable impression of what Durkheim called collective effervescence. Twenty million or 200 million people feel themselves as a force.

The other strand of environmentalism, however, was the emergence of social ecology as a radical and distinctly fringe cause. In the United States, the key figure in this development was the Brooklyn-born

Murray Bookchin (1921–2006), a former Stalinist turned anarchist, who on finding even anarchism too confining, broke off to become a party of one. He made his first foray into environmentalism in 1962 with the publication of a book titled *Our Synthetic Environment*, which stands in interesting contrast to another book published six months later, Rachel Carson's *Silent Spring*. Carson was a biologist and author of poetic evocations of the natural world, such as her best-selling *The Sea Around Us* (1951). The title *Silent Spring* evokes the disappearance of song birds in the suburbs due to the use of pesticides. The book was a phenomenal success. The American public, seemingly unmoved by Bookchin's warnings that industrial development was cocooning society in chemicals, became teary at Carson's image of no more orioles singing in the orchard.

Bookchin took heed. His breakthrough idea was to hitch his radical political agenda to the American love of nature. He enunciated this in a 1964 essay, "Ecology and Revolutionary Thought," where he introduced the term "social ecology."[2] Bookchin went on energetically promoting his brand of social ecology until his death in 2006, spending much of his effort in intramural controversies with other fringe groups on the left. He despised Marxists as promoting new forms of hierarchy; he attacked the Students for a Democratic Society as covertly totalitarian; and he ended up in heated opposition to the group called Earth First! which, in its headier moments, promoted voluntary human extinction as the best way to assist the planet. For Bookchin, the goal of environmentalism was not to save the planet *per se* but to create ultraegalitarian human societies. He pictured himself as a rational thinker in command of "dialectical reason." In his major synthesis, *The Philosophy of Social Ecology: Essays on Dialectical Naturalism*, he asserts that his dialectic is aimed at investigating, "a highly graded, richly entelechal, logically eductive, and self-directive process of unfolding toward ever-greater differentiation, wholeness, and adequacy insofar as each potentiality is fully actualized given a specific range of development."[3] I have no idea what this means, but I am not deceived by it into thinking that Bookchin was an especially intelligent observer. He often evoked Native American tribes living in egalitarian harmony with each other and with nature as his ideal—inadvertently showing his profound ignorance of ethnography and archaeology.[4] He was clear, however, about his central point. As he put it in "Ecology and Revolutionary Thought":

I submit that an anarchist community would approximate a clearly definable ecosystem—it would be diversified, balanced, and harmonious.[5]

Bookchin had a minor academic career. In the 1970s, he founded and directed an Institute for Social Ecology in Vermont which taught courses in "ecophilosophy," and he also taught at Ramapo College in New Jersey, which implausibly elevated him to full professor of "social theory." Along the way, he wrote a historical study, *The Spanish Anarchists*.[6] But it would be safe to say that Bookchinism did not enter into the mainstream of American political or academic thought during his lifetime.

To summarize so far: the sustainability movement arose outside the university. It has achieved worldwide political prominence through international conferences and treaties under the aegis of the United Nations. Within particular nations, sustainability no doubt has many different political trajectories, but it is clearly a force to be reckoned with everywhere. This international concern has a populist basis that can be traced back to agitation in the 1960s over air and water pollution, culminating in Earth Day in 1970. Submerged in that mass movement in favor of cleaning up the environment was another more intellectualized and far more radical critique of Western industrial civilization. This radical critique splintered into many adversarial camps that were nonetheless united by the view that a truly successful environmental movement would have to dismantle competitive capitalism and rid the world of human exploitation. In the view of social ecologists, led by Murray Bookchin, "exploitation" is an undivided reality. Exploiting nature, exploiting biological diversity, and exploiting other human beings are all one and the same. Thus, Cerberus grows his second and third heads. A sustainable world is one in which environmental stewardship is united with social economics and a form of social justice based on radical egalitarianism.

In the mid-1990s, this three-headed conception of sustainability was an extreme rarity in American higher education—perhaps not much more than a provocative idea espoused by a handful of assistant professors hoping to make their mark. As late as the first year of this century, it was no more than a talking point in a few disciplines known for their attraction to edgy ideas. But today, *sustainability* is far and away the dominant ideology in American higher education. I suspect it has similar status in some other nations, especially Britain, but I intend to confine my remarks mainly to the American situation.

Having described my quarry, now I venture on to the more circuitous approach. I have said higher education in the United States is today a troubled institution. The embrace of the sustainability ideology is an instance of that trouble and a symptom of deeper troubles. To track down sustainability, I need to survey that terrain.

Four Horses

I say higher education in the United States is today a troubled institution, but this is, of course, by no means a novel situation. Higher education has been a troubled and troubling institution from ancient times. Partly this is because it contains or attempts to contain several social missions that restlessly pull against one another. Picture four horses with their riders hitched together with long ropes, as each attempts to gallop off in his own direction. The pursuit of truth through systematic inquiry tugs in one direction, while the transmission of cultural and civilizational heritage to a rising generation tugs in another. The practical preparation of students to work in fields that require advanced study strains in still another direction, while the insistence on shaping students' minds and character toward an ideal of the educated person strains toward a different horizon.

These tensions are not accidental features of higher education. They are, rather, central to the enterprise, whether it is Plato's Academy, Cicero's and Quintilian's conceptions of the liberal arts, the medieval university, the Humboldtian research university, the American Bible college, or any of the many other forms into which higher education has been cast. The forms vary in large part by how people have sought to resolve the tensions. Plato elevated philosophy; Cicero denigrated philosophical inquiry for its own sake and urged the practical cultivation of eloquence. Knowledge without eloquence is better than eloquence without knowledge, he argued, but the ideal is to marry the two by making oratory the great goal of education.[7] The research university that emerged in the nineteenth century gave much more emphasis to the discovery of new knowledge. The passage of the Morrill Act in the United States in 1862 created a species of institutions called land-grant universities that attempted to synthesize teaching and research to advance manufacturing and farming, while making scant room for the liberal arts. The bill's primary sponsor, Senator Justin Morrill of Vermont, regarded classical learning as a pernicious nuisance.

But even when the dream of a university's founders was to give clear lead to one of these purposes, the others could be only temporarily diminished. The underlying social realities reasserted themselves. Cicero would not have needed to inveigh against Greek philosophy had not Plato and Aristotle continued to seduce good Romans into this idle pursuit. Land-grant universities in the United States soon found that to secure prestige they would have to compromise their utilitarian curricula by adding the liberal arts. By and large, the history of the university is a history of skirmishes on battle lines such as these. The university as a social institution, as well as almost every particular college and university, perpetually struggles with whether its most needful purpose is discovery of knowledge, cultural transmission, practical training, or personal fulfillment. Universities produce an abundance of rhetoric aimed at vanishing these implicit conflicts into an imaginary harmony, but the tensions are not so easily resolved. They are, in fact, probably irresolvable.

This fourfold schema of purposes that I propose does not easily map on to some other familiar ways of typologizing universities. It does not foreground the division between science and the humanities; it does not emphasize the difference between sacred and secular purposes; and it does not classify universities by their sources of patronage. In my view, all of these are important factors, but they fall into place if you attend first to the question of how a university attempts to yoke together its unruly team of horses: urgent Truth always seeking the better account; vibrant Culture longing to go home; stalwart Utility eager for the day's work; and bright Aspiration ready for the race.

Ben-David and the Equine Analysis of American Higher Ed

I come to this conference, I suspect, as the only participant who encountered the work of Joseph Ben-David for the first time only after I was invited here. I have made some effort to come to terms with his writings, but I have to beg the indulgence of my colleagues here for my much less secure command of Ben-David's sociologies of higher education and science. One of the first things that struck me about his account of the university is his lucid portrayal of higher education as a struggle among advocates for conflicting institutional purposes. Ben-David clearly has his favorites in this struggle. He would like the conditions for creative science to prevail. He believes that transdisciplinary connections in the sciences are profoundly fruitful. He abhors the intrusion of the state and other external bodies bent on turning the university into an instrument to advance their political agenda, and he is especially alert to the danger posed by totalitarian regimes.

He does not find the university anywhere to have achieved a delightful equipoise in which it has found an abiding answer to all its dangers and discontents. Rather, he pictures even the relatively good situations as fragile and vulnerable to disruptive pressures.

This seems to me a sober and fairly compelling picture. But Ben-David made no claims to be a prophet, and I find relatively little in his work that anticipated the troubles that currently beset the American university. If I apply my image of four horses to the American university, the situation is something like this. The discovery of new knowledge is owned now almost exclusively by the natural sciences. They are outwardly healthy and prosperous, and dominate the university in a material sense. But they are inwardly weak. Much of academic scientific research is trivial, and a fair amount of that research is utilitarian work on behalf of commercial ventures. Science has dwindling appeal to American students, though American departments draw many students from abroad. Worst of all, university science has been infected by ideology. Increasingly, it is in the grip of political orthodoxies, in which open questions are settled not by rigorous inquiry and evidence but by enforced conformity miscalled "consensus." More on that later.

The second horse, the one that tugs toward cultural transmission, is on his last legs in American education, and barely exerting any pressure at all. The university has over the last forty years or so progressively defined itself in opposition to Western civilization and American traditions. A student who attends college in the United States hoping to acquire title to this heritage has to move far to the margins to find a small sectarian college or a highly specialized program in a university. Otherwise, he will encounter mostly a banal denunciation of the last 2,500 years of Western history as a story of endless oppression of minority groups by triumphalist elites. A certain culture is indeed transmitted in this fashion, but it is the extraordinarily narrow one that Ben-David called "an entrenched tradition of academic radicalism."

As Ben-David pointed out, the appeal of radical movements for students and young teachers arises in part from the stress of having to "prove themselves through intellectual attainments" and "Joining radical movements reduces this stress by conferring on them the status of intellectuals without requiring proof of scholarship."[8] Ben-David's somewhat cynical apercu is accompanied by a footnote in which he further distinguishes those who arrive at radicalism though insecurity and those who stay there out of opportunism. He seems not to have held an especially elevated view of campus leftists. The third horse in his traces is that of practical training. Why American higher education continues to enjoy a good reputation in this area is hard to explain. American colleges and universities fail to graduate within six years more than half the students who enroll as freshmen. And those who do graduate frequently fail to acquire much in the way of knowledge or practical skill. A bachelor's degree from an ordinary American university or even an elite university is no longer taken at face value to signify that a person understands the basics of any field. At best, the degree is an invitation to inquire further. That is to say, the horse of practical training has devolved into the mule of sterile credentialism. There is a very strong utilitarian streak in the American character, and students who have once set their sights on learning how to do something will generally do it. But the American university has only faint and intermittent interest in helping.

Rites of Passage

The fourth horse I characterized in several different ways. I referred to it first as the university's insistence on shaping students' minds and character toward an ideal of the educated person; but then I summarized it as the pursuit of personal fulfillment, and finally named it as the race horse Aspiration. These, of course, are not always the same thing, but what I mean to evoke is that sense in which the focus of education is on becoming a better person. Higher education has often been conceived not merely as *acquisition* of knowledge, skill, or culture, but as internal change. Sometimes, this is conceptualized as character building; sometimes as liberation, or growth, or self-actualization. A comparative anthropology could surely be written on how societies thematize this particular rite of passage, but in this essay I mean only to mark it as one of the perennial characteristics of higher education.

In the United States, this has become in an odd way the most powerful part of undergraduate education. The emphasis on science, research, and the discovery of knowledge is far more conspicuous, but the actual lives of most college students have little to do with such science or discovery and quite a lot to do with a quest for a sort of moral renovation and legitimacy. Half a century ago, American colleges routinely expressed this with the term I used a moment ago: character building. That term would now strike most Americans as a quaint expression or an outlook to be found exclusively in religious colleges. But if the term "character building" is archaic, the underlying conception flourishes in new dress. American higher education is now awash in a rhetoric of "personal transformation," "educating the whole person," and, more vaguely, "leadership."

The switch from the old term to the new rhetoric registers several cultural and social shifts. "Character building" belonged to an age in which American higher education was for the few, and the aspiring elite to whom it catered needed to acquire the habits and outlooks suitable to a managerial class. The "character" that was meant to be built by a college education was rooted in honesty, generosity, self-discipline, confidence, respect for others, and public-spiritedness. A college education was meant to enhance these qualities, which were assumed to be already present in students but in need of further development.

College in America is now conceived as a project not for the few but for the many. It is a system of mass higher education in which one of the principal discontents is that it has not expanded nearly far enough. U.S. presidents and national politicians continually call for sending higher and higher percentages of high school graduates to college with no regard for the ability of these students to meet the intellectual rigors of a college curriculum. Clearly an emphasis on the kind of character education meant to cultivate in students the values of a commercially minded social elite was no longer appropriate. The incongruity struck students during the cultural revolution of the 1960s, and American higher education fairly quickly absorbed the message. The *in loco parentis* policies in which universities enforced bland control over student life were abolished. Disciplinary codes were transformed into student-run committees under loose supervision of college officials. Student governments and extracurricular activities became less and less supervised and faculty advisors for such groups receded to titular roles. For a period, student autonomy was, on many campuses, unencumbered by any meaningful adult presence.

This was the nadir of character education in American universities in the 1970s and 1980s. But, as I have said, none of the foundational purposes of the university ever really disappears. If one of these purposes is suppressed or neglected for a time, it inevitably returns with pent-up force. In this case, the force was revealed by a national study carried out by the Carnegie Foundation at the end of the 1980s which captured the deep unhappiness of American college students with the results of the abandonment of character education.[9] By far the chief complaint of college students recorded in that study was the lack of "campus community." College life was experienced as isolating and fragmented. The retreat from the effort of shaping students in favor of leaving them alone to make what they could of their freedom turned out to produce a sense of aimlessness.

The Carnegie Foundation report also came at the end of the decade in which American universities had enunciated a new ideal: the concept that education would be advanced by "celebrating diversity." The term, "diversity," was taken from an opinion in a U.S. Supreme Court case in 1978, in which the court ruled that a white student had been unfairly excluded from admission to medical school because of that school's use of a race-based set aside. One of the judges speculated that, had the university justified its racial quota by invoking the educational benefits of racial diversity in the classroom, he would have upheld the university's decision. Soon, colleges and universities began to rename their offices of affirmative action as "diversity offices." Though it was initially a term without much content, it caught fire. The supposed academic good of diversity became the rationale for a wide variety of campus programs that focused on fostering minority group solidarity and justifying academic double standards not just in admissions but also in faculty appointments, the curriculum, grading, and more. Diversity was the intellectual gloss put on identity group politics. Added to a campus social environment in which the bonds of shared community were already slack, diversity was a recipe for replacing the campus as the site of shared aspiration with the campus as a site of mutual antagonism among ethnic-based cohorts. Diversity evolved into a doctrine that encouraged identity groups to hug their resentments and to "study" the history of injustices and grievances they were heir to, while at the same time commanding the community at large to "celebrate diversity" as an enriching experience.[10]

The death of *in loco parentis* as the practical basis of character education and the rise of diversity as moral validation of ethnic hostility set the stage for the Carnegie Foundation's finding that students now yearned to be part of a larger solidarity. But what might that larger community be, especially if it needed to reconcile itself with the adults' fixation on emphasizing separatist identity groups?

The answer was not long in coming. In 1994, the American College Personnel Association, a body that represents the administrators and staff who work in extracurricular activities on campus, issued a report titled *The Student Learning Imperative: Implications for Student Affairs.*[11] It crystallized a new approach to an old university function. The staff members whose job it had been to maintain suitable and safe dormitories and to preside over student clubs and extracurricular events declared in *The Student Learning Imperative* that their real task was something much more important and more elevated. They were entrusted—or were ready to entrust themselves—with a dimension of higher education larger than the curriculum itself.

In one sense, *The Student Learning Imperative* documents the efforts of a low level group of service employees to claim professional status. The manifesto is a way of asserting, *We are just like professors. We too are responsible for educating students.* This envy is hardly hidden. At one point the document declares:

> Student affairs professionals are educators who share responsibility with faculty, academic administrators, other staff, and students themselves for creating the conditions under which students are likely to expend time and energy in educationally-purposeful activities.

But if student affairs staff are educators, what is the content of what they teach? This seems, at first, a bit vague. We are told these professionals:

> . . . endorse talent development as the over-arching goal of undergraduate education; that is, the college experience should raise students' aspirations and contribute to the development of skills and competencies that enable them to live productive, satisfying lives after college.

The Student Learning Imperative, however, does yield a curriculum of sorts, dubbed by some critics, "the shadow university."[12] It is a curriculum that aims to restore a version of character education and community building. Its desiderata, as presented in *The Student Learning Imperative*, include various "hallmarks of a college educated person" such as "understanding and appreciation of human differences" and "conflict resolution," and "a coherent integrated sense of identity, self-esteem, confidence, integrity, aesthetic sensibilities, and civic responsibility." These anodyne goals, however, are said to be poorly achieved in an institution that splits "affective and cognitive development." Cognitive development is the stuff of classrooms and traditional academic learning. By its nature, it compartmentalizes and fragments, and leaves students with what the report writers call "disjointed, unconnected experiences." The larger, "holistic philosophy of learning" espoused by the student affairs professionals, by contrast, heals these rifts, bridges "organizational boundaries," promotes "personal development," and creates the "psycho-social" conditions in which students thrive. This holistic philosophy encompasses what are called "process values" such as ethnic diversity, gender balance, equity, and justice.

The aggrandizement of this category of university staff thus emerges as a vision of the bureaucrats as both overseeing a therapeutic milieu and advancing a social agenda which includes a version of the social goods promoted by the political left in the United States.

Fifteen years later, *The Student Learning Imperative* remains in active circulation, promoted by its sponsoring organization, the American College Personnel Association, but it has, of course, long since been elaborated in other pronouncements and has seeded its own body of quasi-empirical studies intended to make good on the idea that "student affairs" is an academic discipline in its own right. But I do not intend to follow this thread much further. There is no need to because, lo and behold, there at the edge of the clearing, beside that dark stream, stands our quarry: sustainability itself!

Sustainability in the Dorms

In October 2007, I learned from a faculty member at the University of Delaware about a compulsory program for that institution's undergraduate students run through that institution's residence halls. The explicit aim of the program was to "transform" students by using a variety of consciousness-raising techniques. Toward this end, for example, groups of students would be brought to a meeting room and be told to stand in one place if they supported gay marriage and in a different place if they opposed gay marriage. No one was permitted to be neutral or to decline to divulge his views. Another activity consisted of splitting students into small groups that were assigned the task of compiling as many racial and ethnic epithets as they could think of.[13] Another exercise consisted of segregating white students and publicly ridiculing them for their alleged privileged position in American society.

Much of the University of Delaware program was entrusted to older students, hired as "residence assistants," or RAs, who were specifically trained for the task. The training included learning some unusual definitions for common words that the RAs were then supposed to teach to the other students.[14] One word provided with such a definition was "racist," which according to the RAs' training manual is:

> One who is both privileged and socialized on the basis of race by a white supremacist (racist) system. The term applies to all white people (i.e. people of European descent) living in the United states, regardless of class, gender, religion, culture or sexuality. By this definition, people of color cannot be racists, because as peoples within the U.S. system, they do not have the power to back up their prejudices, hostilities or acts of discrimination.

The Delaware program also laid out a number of "competencies" that all students were expected to acquire through this dorm-based training. They included:

> Each student will recognize that systemic oppression exists in our society.
>
> Each student will recognize the benefits of dismantling systems of oppression.

The program also entailed the goal of inculcating in students principles of good "citizenship," where the word "citizenship," like the word "racist," was freighted with unusual meanings. Good citizenship, for example, included support for "fair trade," "domestic partnerships," "gender equity," and "rights of indigenous peoples."

Coercive techniques included placing a single student in the center of a room ringed by other students who were encouraged to pepper him with hostile questions about his social attitudes. RAs were also required to meet individually with students to obtain answers to a formal questionnaire. One of the questions was, "When did you discover your sexual identity?" Students who resisted were written up in a report that became part of their permanent dossiers.

We know all this because a controversy erupted over these techniques; internal documents from the university's office of residence life became public, and numerous students came forward with accounts of their experiences. Under pressure, the university suspended the program, only to bring it back a year later shorn of some of its compulsory elements, but still in many respects compulsory.

Those hearing of this University of Delaware program for the first time often respond with incredulity that a university devoted, in principle, to free intellectual inquiry, would contrive a structure to force crude political propaganda on its undergraduate students.

Incredulity as well that those students could be systematically placed in situations where they would be intimidated, humiliated, and browbeaten into both false confessions of guilt and equally false professions of loyalty to a political credo chosen by university authorities. This sounds like something that might occur in Revolutionary Iran or some twentieth-century totalitarian dictatorship, not a public university in a constitutional republic with a Bill of Rights.

So how did the University of Delaware convince itself that this was a proper use of its educational authority? It launched and justified this program in the name of "sustainability."

When I first saw this connection presented in the University of Delaware's own documents, I found it baffling. The word "sustainability" to me resonated only as one of the talking points of environmentalists and meant something along the lines that we should strive to regulate human production and consumption so that these activities would not destroy their basis. It is a word suffused with a sense of responsibility toward natural resources and future generations. I could see glimpses of this kind of sustainability in the Delaware program which, for example, included among its other pedagogical goals the demand that:

> Each student will be able to utilize their [sic] knowledge of sustainability to change their [sic] daily habits and consumer mentality.

Delaware students were also pressured to take a pledge to reduce their "carbon footprints" by 20 percent. These struck me as inappropriate demands for a university to make, and rather short of a reason to call the whole program, with its focus on racial attitudes, sexual identity, and societal oppression, a *sustainability* program.

I had much to learn. And perhaps I still do. But I now understand that radical environmentalism and other forms of radical social egalitarianism had been married by utopian thinkers such as Bookchin as long ago as the 1960s. The combination of environmentalism with radicalism of the left, however, remained on the social fringes for decades and had no meaningful institutional base within the universities. This presents a key historical question: how did this fringe political movement become a dominant force in American universities? This absorption of the sustainability ideology into higher education is far from complete, so this is not a matter of evaluating in retrospect a finished process of institutionalization. But one phase of that transition has been finished. Almost no one in American higher education now sees the sustainability movement as an outrageous muddling of categories; a conflating of environmental concerns with rightly separate economic, social, and political matters; or a fusion of several radical anti-Western themes. To the contrary, sustainability now names a mainstream and respectable enterprise. It does so even though, in essentials, it remains a muddling of categories, a conflating of concerns, and a fusion of disparate radical anti-Western themes.

In 2006, a year before the controversy erupted at Delaware, Kathleen Kerr, the head of the dorm-based indoctrination program, enunciated her view of how sustainability could wrap together environmental advocacy with elements such as the fight against racism. She made a presentation at a conference on "Tools for Social Justice" for student affairs staff and spoke on "Sustainable Development: Toward Healthy Environments, Economic Strength, and Social Justice."[15] On this occasion, she declared that it is a "myth" that "Sustainability is mostly about the environment," and laid out a lengthy list of the "social justice" aspects of sustainability. These included "environmental racism, fair trade, living wages, domestic partnerships, rights of indigenous people, gender equity, affirmative action, and multicultural competence." I have only an outline of her talk, but pieces of it reappear in other documents in fuller form. For example, one of the University of Delaware planning documents for the sustainability program, dated 2007, lays out this explanation:

> Sustainability provides a mechanism to take a comprehensive look at the interconnections that exist between ecological, economic and equity issues such as global warming, pollution, health and poverty and work towards lasting solutions (Edwards, 2005). While students will not be prepared to, nor are they expected to tackle these complex topics during their freshman year, it is imperative that they begin to develop a value system that considers how their actions contribute to the further augmentation of these issues. As such, sustainability provides a viable conduit for citizenship education and the development of a particular values system.[16]

Note in this wording—which I believe to be Kerr's—that the author presents sustainability not as an end virtuous in its own right. She may believe that it is, but her argument here is that sustainability is a handy means to get other desirable things done. She begins by saying, "Sustainability provides a mechanism" for bundling together otherwise disparate issues and she ends by saying, "sustainability provides a viable conduit for citizenship education."

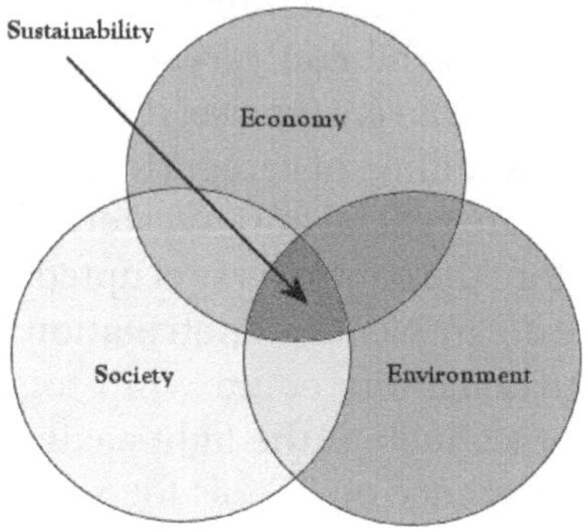

Sustainability in this version of the ideology is the delivery truck. The contents are a collection of issues that, at first glance, have little else in common. The sustainability movement in fact is a backhanded way of acknowledging this makeshift unity. It has adopted as a symbol a Venn diagram of three overlapping circles that look vaguely like the five interlaced rings of the Olympic Games. But in the case of sustainability, these circles are labeled: the environment, the economy, and social justice. The area where the three intersect is typically labeled "sustainable development." In some versions, the intersections of two but not three of the circles are labeled "compromise." This is the triad that I have likened to the three heads of Cerberus.

A diagram can make an idea seemingly vivid without making it actually clear. In exactly what sense do environmental, economic, and social justice issues converge to create "sustainable development?" The diagram offers no answers any more than if one drew three circles labeled God, science, and art and cited their intersection as defining the province of angels. But the overlapping circles of sustainability are a common sight on American campuses: visual shorthand for commitment to the movement and to the assertion that radical versions of economics, social justice, and environmentalism converge on a single project.

I have mentioned one way in which this overlap can be brought into the compass of a social theory: Bookchin's way. If we view "exploitation" as a unitary category that applies equally to inert materials, living things, and people, and if we view a "sustainable society" as one that eradicates exploitation, then there is a rough unity of environmentalism, socialist economics, and a radically egalitarian version of social justice. Bookchin's synthesis, however, is not the only possible approach. Kathleen Kerr, the Delaware residence life savant, evokes the UN Decade of Education for Sustainable Development, not Bookchinism, as her point of departure. She claims an ideal rooted not in the anarchists of the Spanish civil war or the self-sustaining American aborigine hunter/gatherers, but in the notion of "global citizenship." It is difficult to extract from her writing a coherent account of how environmentalism, economics, and social justice are meant to converge, but the central idea seems to be that the individual student is the point of convergence. The aim of the Delaware program is to liberate the student from the prejudices that keep him bound to the rapacious social system in which he grew up. He attains global citizenship by progressively divesting himself of bad social attitudes and replacing them with good ones. The "bad" social attitude he has acquired growing up in America consists of the value he places on his "independence," his "materialism," and his willingness to treat the "market as master." Liberation for him consists of coming to value "interdependence," "human satisfaction," and the "market as servant."[17]

My encounter with Kerr's ideas and the University of Delaware program prompted me to ask, "How many University of Delaware-style residence life programs are there in the United States?" To find out, I hired a new staff member, Tom Wood, at my organization, The National Association of Scholars, to pursue a systematic inquiry titled, *How Many Delawares*? After all, Kathleen Kerr, a key figure in the American College Personnel Association (ACPA) and frequent lecturer on sustainability programs, had been promoting the Delaware model for some years. And she was clearly not alone in her enthusiasm. Her organization, ACPA, published its collective manifesto on the topic in 2008, *Toward a Sustainable Future: The Role of Students Affairs in Creating Healthy Environments, Social Justice, and Strong Economies.*[18] Here she is one of thirteen authors, representing residence life or related programs at ten universities. (The document declares: "Each of us must adopt sustainability as a new personal and professional paradigm—a new way of thinking about our stewardship of our environmental, human, and economic resources.")

Tom Wood found nothing else quite like the University of Delaware's dorm-based compulsory indoctrination program but turned up hundreds of cases of what he called "The Communitarian ResLife

Movement" in American universities.[19] But he did document that "sustainability" had become a major preoccupation of student affairs personnel in American universities and that the term was used almost everywhere as a rubric for dealing with issues such as racism, sexism, and matters of personal identity. Tom Wood argues, as the title of his report suggests, that the concept that holds this together is a version of communitarianism. The link, he says, is that both communitarianism and sustainability view "societies based on individualism to be unsustainable in the long term."

I have some reservations about how well this label fits the overall phenomenon of the sustainability movement. For one thing, sustainability is linked to an approach that valorizes identity groups that base themselves on grievance and mythologized suffering, and in that sense sustainability is an agonistic ideology. It also fosters intergenerational resentments. One of the ways in which sustainability enters education is by pointing an accusatory finger at parents and the older generations, who are depicted as wantonly and ignorantly putting the world at risk. Communitarianism, as it is usually understood, emphasizes the moral bonds across ethnic groups and between the generations. But Tom Wood sees an "alliance" between multiculturalism and left-wing communitarianism based on the fight against entrenched inequalities. He also suggests that the sustainability movement's emphasis on protecting biological diversity conjures a principle that can be applied to efforts to sustain ethnic and cultural diversity.

In any case, Tom Wood's long series of reports to the National Association of Scholars on sustainability programs is rich in detail and goes far toward filling in the picture of how the sustainability movement became institutionalized in American universities through the back door of student affairs bureaucracies. The ideals of sustainability were firmly planted in higher education without ever coming before university faculties for serious consideration or debate. Not that the faculty necessarily minded.

Sustainability in Mind

I hypothesize that sustainability found its way on campus through the efforts of a marginal caste of bureaucrats hoping to improve their status and to gain a larger share of institutional resources. Some of the evidence to support this hypothesis lies in the behavior of these bureaucrats. The sustainability program at the University of Delaware was, among other things, an occasion for the bureaucrats to mine the responses of students to the various initiatives for data that could feed a research publication program. The associate director of residence life set out a rubric in 2006 for measuring how well students achieved the "learning goals" and the twelve "competencies" set forth as part of "citizenship." These were to add up to publishable studies on matters such as students' success in learning "the skill necessary to be a change agent," and their ability to "demonstrate civic engagement toward the development of a sustainable society." The sheer silliness of this sort of thing seems to have eluded all the participants. The enterprise, attested by hundreds of pages of documents, is suffused with a tone of earnest ambition.

But I must acknowledge that my hypothesis of how sustainability entered the university and took root is still only hypothesis. Sustainability was also gaining ground on campus beginning in the 1990s from other directions. There was at least one foundation created in the early 1990s dedicated to bringing the sustainability doctrine to campus. I will come to that a little later. Academic books and papers on sustainability are now abundant.[20] As the field becomes a site for new scholarship, contributors scour the past for useful precedents. Bookchin has been rediscovered and many old writers dusted off. The dour H.G. Wells of *The Time Machine*, for example, has been dusted off. Here he is quoted in the foreword of the *Sustainability Revolution* blessing the doomsayers by pronouncing in 1946, "The end of everything we call life is close at hand and cannot be evaded."[21] The author of *The Sustainability Revolution*, Andres R. Edwards, reaches still further back to credit the New England transcendentalists, including Thoreau and Emerson, with the foundations of the movement. Pedigrees confer prestige and legitimacy, but they are not necessarily good history or sociology. Edwards proclaims a sustainability revolution, but benchmarks its progress mainly by the series of UN conventions that I cited earlier. He also offers an account of some of the numerous activists and advocacy groups at work in the field since the 1970s, but very little points the way toward an ideological invasion of the university. He cites one work, however, that seems to be a touchstone for sustainability in academe, David Orr's 1993 volume, *Earth in Mind: On Education, Environment, and the Human Prospect.*[22]

Orr, a professor of environmental studies and politics at Oberlin College, is an advocate of teaching children the value of "the responsibilities of informed citizenship," which to him means environmental awareness coupled with skepticism toward individualism. In a preface to a new edition of *Earth in Mind* in 2004, he allows that "against considerable odds" environmental education "is becoming well established" in colleges and universities around the world. In the first chapter of the book, "What Is Education For?" Orr depicts modern education as riddled by destructive myths, which include the myth that ignorance can be overcome by learning; the myth that knowledge and technology can help us "manage planet earth"; and the myth that "knowledge is increasing." It seems reasonably clear that Orr is not headed in the direction of Humboldt's university. Instead, Orr values intellectual modesty, shaping ourselves "to fit a finite planet," and holding on to vernacular knowledge and folk culture. He dismisses much that the research university aims at as mere "cleverness," and holds that "true intelligence is long range and aims toward wholeness." Another myth he would debunk is the utilitarian aim of preparing students for "upward mobility and success." Would I be engaged in parody if I drew from this that sustainability has no room for worldly success? No, Orr says it himself. "The plain fact is the world does not need more successful people."

It is fairly clear what Orr is against in higher education; a little less clear what he favors instead. He is against "education that alienates us from life in the name of human domination, fragments instead of unifies, overemphasizes success and career, separates feeling from intellect and the practical from the theoretical, and unleashes on the world minds ignorant of their own ignorance." The alternative? Orr suggests that we need to "rethink education" to focus on "issues on human survival," and that "All education is environmental education." He is an advocate of connectedness, rootedness, and above all alarm. He would refocus education away from knowledge and toward "intellectual emphasis, loyalties, affections, and convictions"—a litany that he borrows from the mid-century American naturalist Aldo Leopold, author of *A Sand County Almanac (1949).*

Orr's book was clearly influential, although I am not far along enough in my study of this movement to venture a guess as to how much of a following he gained in the university. He does, however, turn up with considerable frequency in contemporary writings by the people I have come to call academic sustainatopians. Earlier this year, I published an article commenting on a paper in the prestigious peer-reviewed journal, *Current Anthropology*.[23] The paper, "Reason and Reenchantment in Cultural Change: Sustainability in Higher

Education," by Peggy F. Barlett, the Goodrich C. White Professor of Anthropology at Emory University, struck me as an astonishing event.[24] Here was a journal, quite plainly dedicated to scientific inquiry and the norms of rigorous scholarship, presenting a straightforward claim that, in light of the existential peril facing the planet and concomitant need to advance the sustainability project in higher education, we need to abandon disinterested inquiry in favor of such things as helping students achieve "heightened sensory awareness." Barlett draws on a growing body of writers who want to "expand the scientific paradigm of an objective relationship with the natural world (based on science and the use of reason) to include a more personal reconnection with the living earth." In Barlett's view, reason and rationality are part of the problem. They contribute to the sense of fragmentation that stands in the way of "enchantment" which "involves dimensions of fascination, mystery, delight, and awe," that science knows not of.

Barlett's essay is a polemic against the university which, in her view, is still mired in its old ways and not sufficiently enthusiastic about sustainability. Even business and government are ahead of higher education, which is "inherently conservative and highly fragmented." By now, the terms of this criticism should be thoroughly familiar. Orr, whom Barlett cites, sees the university in this light, as did the authors of *The Learning Imperative*, and Kathleen Kerr. Scratch the world Tom Wood called the "Communitarian ResLife Movement," and it will bleed scorn for the disciplinary divisions of the university which stand in the way of a "whole person pedagogy." Venture into the world of Peggy Barlett and prepare to be buffeted on all sides by writers who view the disciplines as impediments to a salvific vision of the dissolution of the individual into the cosmic unity of things.

Sustainability from the Corner Office

Sustainability is an ideology of the left promoted with one eye on career advancement by student affairs bureaucrats, and sustainability is a quasi-religious doctrine that has gained a Bacchae-like following of academics intent on uniting everyone into the cult. But sustainability in the American university is also several more things. It has become a new type of academic program. Arizona State University was among the first to create a degree in sustainability. There are now fifteen *degree* programs in sustainability at eleven universities, and according to AASHE, "more than 66 sustainability-focused academic programs were created last year."[25] I have this information from the Association for the Advancement of Sustainability in Higher Education or AASHE a body founded in 2005 "to help coordinate and strengthen campus sustainability efforts." It now has 660 colleges and universities that are members and holds workshops and conferences throughout the year. AASHE is not alone. There are also the U.S. Partnership for Education for Sustainable Development, the Higher Education Associations Sustainability Consortium, and the National Association of College and University Business Officers (NACUBO) which has its own dedicated section on sustainability emphasizing the need for colleges and universities to reduce or neutralize their carbon emissions. Among other things, NACUBO promotes the *American College and University Presidents [sic] Climate Commitment,* which has so far garnered the signatures of 650 college and university presidents who all want to cut carbon emissions and "re-stabilize earth's climate."[26]

Though not as floridly mystical as Professor Barlett's article in *Current Anthropology,* the Presidents Climate Commitment also has a statement of faith. It begins:

> We, the undersigned presidents and chancellors of colleges and universities, are deeply concerned about the unprecedented scale and speed of global warming and its potential for large-scale, adverse health, social, economic and ecological effects. We recognize the scientific consensus that global warming is real and is largely being caused by humans.

I realize I venture onto thin ice here, but I know of no real scientific "consensus" that global warming is largely caused by humans. I know of many scientists who say it is, but also quite a few who doubt that proposition. An 850-page tome arrived on my desk some time ago, titled *Climate Change Reconsidered,* which presents some sixty articles by scientists from around the world who dispute the notion that global warming is "largely caused by humans."[27] These dissenters may be mistaken, but their existence gives the lie to the word "consensus" in the claim that there is a scientific consensus for this proposition.

Also, in my position as head of the National Association of Scholars (NAS), I hear with some frequency from well-established university-based scientists who report on the tactics of intimidation and reprisal that accompany the efforts to manufacture scientific consensus where there is, in fact, no such thing. Much of the problem arises from the nature of climatology, which at this stage of its development relies a great deal on speculative modeling. One has to view these models with great intellectual generosity to invest their projections with credibility. As a result, debates over global warming and climate change inevitably turn into metadebates over where the burden of proof should lie. Should we uphold the hypothesis of man-made global warming pending irresistible contrary evidence? Or should we refrain from definite conclusions pending the emergence of scientific models that can be substantiated?

In any case, the Presidents Climate Commitment has a certain odd quality to it. Well-established scientific facts, the sort that we would unhesitatingly describe as backed by consensus, do not need testimonials from college presidents, few of whom are themselves scientists, and fewer still scientists in the relevant disciplines. We do not get a Presidents Gravity Commitment or a Presidents Germ Theory of Disease Commitment. The very purpose of making such a declaration is ideological. It is meant to silence unwelcome debate.

My colleague at NAS, Ashley Thorne, has been tracking the Presidents Climate Commitment and has picked up some of its dubious maneuvers.[28] The statement avows that campuses that integrate sustainability into their curricula, "will better serve their students and meet their social mandate to help create a thriving, ethical and civil society." That is potentially the same mandate we saw at the University of Delaware. The Climate Commitment is sponsored by AASHE as well as two other organizations, and AASHE has supplemented it with a wiki offering college presidents further tips on how to make the commitment "high profile" and permanent.[29] It calls for making sustainability part of the education of every student, empowering "student environmental activism," promoting "residential environmental education initiatives" and, perhaps most ominously, imposing a "climate action litmus test" on the hiring of all future campus leaders.

The Presidents Climate Commitment has been staffed and supported by three organizations: AASHE, and also ecoAmerica and Second Nature. EcoAmerica is an organization, not specifically focused on higher education, that espouses a startlingly frank program of attempting, by means of "psychographic research" and "engagement marketing," to "shift personal and civic choices of environmentally agonistic Americans." It is, in other words, devoted to propaganda and manipulation. Second Nature was founded in 1993 in Boston with an explicit goal:

> to accelerate movement toward a sustainable future by serving and supporting senior college and university leaders in making healthy, just, and sustainable living the foundation of all learning and practice in higher education.

Second Nature claims to have worked with over 4,000 faculty members and administrators at more than 500 colleges and universities, and it began with both key political connections and substantial financial support, its founders including Senator John F. Kerry and his wife, the heiress Teresa Heinz Kerry.[30] Second Nature's multimillion dollar advocacy aimed at the leadership of higher education seems an important part of the story of sustainability's rise to the position of dominant ideology in the American university—but this is a thread I have not had the opportunity to investigate.

Sustainatopian Sadness and Exuberance

What kind of thing is the sustainability movement? I take it as an illiberal and strongly anti-intellectual ideology. It finds its place in the university mainly as a vast elaboration of only one of the four perennial purposes of higher education: the goal of shaping students' aspirations toward an ideal of the educated person. This is the "imperative" in *The Student Learning Imperative*: an attempt by means of therapeutic techniques, aversive intervention, propagandistic repetition, soothing emotional appeal, flattery, intimidation, censoring alterative views, and appeals to authority to establish a campus orthodoxy. The sustainability movement promotes a few simple ideas. It is by no means hard to grasp, though it can be very hard to please. It asks of its adherents a great deal of self-denial, while urging them to promote forced denial on others. It has its own Puritanical fervor. We are all sinners against sustainability. We can only atone. One form of atonement that is called for is intellectual abasement. Expressing doubt about the vaunted "consensus" is further evidence of one's fallen nature. The global warming skeptic, the student who fails to recycle, or fails to recycle enough, the student whose carbon footprint is too large—all these confess by their actions that they still cling to the illusion of individuality and have yet to submit to their interdependence or realize the need for global citizenship.

How does this ideology sit with a university that is still, in many practical ways, a citadel of science? It sits poorly. The sustainability ideology actively discourages some students who might seek a scientific vocation, and it bends others to a notion of science shorn of the discipline of intellectual reserve. Sustainability in the sciences is ensnared in a web of false assurances. I refer not just to global warming but to the image of nature as a complex self-perpetuating system that, absent human interference, retains a destined order. We know rationally and scientifically nature is not like that. Its order includes disaster, struggle, competition, blind-turns, and extinctions. But sustainability teaches a different insight. It invites an image of nature in which hierarchy and exploitation vanish. And by urging us not to overcome the alienation of man from nature, it subtracts one of the basic conceptual steps that gave rise to scientific inquiry.

No doubt science will survive this assault, in no small part because the sustainatopians are by and large too facile to sustain their own moonbeam movement. But they can accomplish considerable harm before their romance with images of eco-apocalypse fades out.

When I say sustainability is illiberal and strongly anti-intellectual ideology, I am also referring to its affinity with the radical left. Clearly only a few promoters of sustainability are Bookchinites, but there remains a question of why concern with a clean and healthy environment has become so entangled with identity politics, which at the present moment tend to be leftist. Tom Wood says the link is mutual hostility to individualism. But I would add that concern for the environment is compatible with almost any kind of political system and a great variety of economic systems. In the United States, the Clean Water Act was a project of President Richard Nixon. For a long time, the leading lights of conservation in America were conservative Republicans. American capitalists have often grumbled over particular environmental enactments but have generally turned into enthusiastic supporters of the broad goal of maintaining a clean and healthy environment. On its face, sustainability could be yoked equally well to a conservative ideology.

Rather than a natural and easy fit between environmentalism, anticapitalist economics, and radical egalitarianism, there are natural and deep fissures among all three. Socialist economies have proven themselves over and over again to be far more rapacious despoilers of the environment than free-market economies. Societies with strong legal traditions that support private property rights have substantially better records of protecting the common environment. But the sustainability movement, as it has emerged, spurns these lessons. It is antidemocratic at its core and deeply invested in the idea that its leaders, through their empathetic connection with the "needs" of the Earth or the biosphere, possess a moral authority that trumps the rights of ordinary people.

I began this essay by saying the American university is a troubled institution and by promising an account of one instance of its troubled nature. I have suggested that the trouble represented by sustainability arose initially because the university made itself vulnerable when it backed away from its old commitment to shape the character of students toward a positive ideal of social maturity. That left among students a longing for the personally transformative side of education. It also created an opportunity for an ambitious class of bureaucrats to step in.

As far as I can tell, it was happenstance that the sustainability ideology came to hand as the actual content of the effort to corral students into residence-life-based programs that aimed at promoting politically progressive causes. When *The Student Learning Imperative* came out in 1994, it had no mention of environmentalism or sustainability. But sustainability proved to be the perfect rubric for bundling together the moral fervor of environmentalism with the seething discontents of identity politics. It also received a well-funded nudge from Second Nature, an advocacy organization intent on bringing sustainability to campus from the top down by means of appealing to campus executives.

How much peril does the sustainability movement pose for the university? The sustainability movement is heir to a forty-year-long effort by the left to make open political advocacy a legitimate part of the college curriculum. The same long-term development has seen an untiring effort to diminish the authority of rational inquiry that attempts to minimize bias and to strive for objective truth. This is not the place to reprise that history, except to say that corrosive skepticism toward traditional scholarly standards combined with epistemological twists aimed at elevating the status of group-think have gained purchase in the humanities and social sciences, and have begun to make headway in the sciences too. The sustainability dogma could not have gotten as far as it has were it not for this already achieved assault on core academic standards. The particular peril posed by sustainability is that it gives a warrant to drive political advocacy into every last corner of the institution. Nothing—not the privacy of the dorm, not the menu in the cafeteria, not the selection of the English 101 textbook, not the grant proposal of the chemistry professor, not the appointment of the next dean, not the architecture of the next building—nothing is exempt. The oddity of this is worth remarking. The institution that once took as its deepest purpose the fostering of free minds is rapidly becoming one of the intellectually least free places in Western society. And it is achieving this in a spirit of exuberant self-congratulation.

A word or two more on the sciences, since the conditions of their flourishing was so much a part of Ben-David's account of the university. At the end of *The Scientist's Role in Society*, Ben-David described the "conditions under which the pursuit of science without recurrent moral crises has been possible." In his view, these include a political ambience that allows "social experimentation and pluralism," and a way for institutions to change "without recourse to violence." He seems to have in mind a liberal social order, and he warns that without the kind of flexibility such societies afford, science is destined to encounter "waves of scientific enthusiasm . . . followed by vogues of antiscientism, romantic irrationality, and even antinomianism threatening the very existence of science."[31]

The threat Ben-David had in mind seems pretty clearly external to the university. When he published *The Scientist's Role in Society* in 1971, science in the United States was flourishing, and it would have been hard to see much warrant for the picture of waves of scientific enthusiasm followed by antiscience and irrationality. We are, however, pretty close to that now. The sustainability movement makes crashingly large scientific claims and mixes them with great swales of exuberant irrationality. Western society seems once again gripped by a vogue of antiscientism and romantic irrationality. But this manic-depressive moment in the life of the university was not thrust on the university by an external political movement. It was nurtured and brought to fruition by the university itself.

Notes

1. World Commission on Environment and Development, *Our Common Future* (New York: The United Nations, 1987), http://www.un-documents.net/wced-ocf.htm (accessed February 27, 2012).
2. Murray Bookchin, "Ecology and Revolutionary Thought," Originally published under the pseudonym Lewis Herber in Bookchin's newsletter *Comment* in 1964. http://dwardmac.pitzer.edu/anarchist_archives/bookchin/ecologyandrev.html (accessed February 27, 2012).

3. Murray Bookchin, *The Philosophy of Social Ecology: Essays on Dialectical Naturalism* (New York: Black Rose Books, 1990), 39.
4. Richard J. Ellis, *The Dark Side of the Left: Illiberal Egalitarianism in America* (Lawrence, KS: University of Kansas Press, 1998). Ellis provides an absorbing account of the conflicts between Bookchin and members of Earth First! In his chapter, "Earth First! And the Misanthropy of Radical Egalitarianism," 228–51. My view of the first attempts to fuse leftist radical with radical environmentalism is greatly indebted to Ellis' scholarship.
5. See note 2.
6. Janet Biehl, "A Short Biography of Murray Bookchin," *Anarchy Archives*, December 18, 1999, http://dwardmac.pitzer.edu/anarchist_archives/bookchin/bio1.html (accessed February 27, 2012).
7. Cicero, *De Oratore* 1.72. Quoted in Bruce A. Kimball, *Orators & Philosophers: A History of the Idea of Liberal Education* (New York: The College Board, 1995), 36.
8. Joseph Ben-David, *Centers of Learning* (New Brunswick, NJ: Transaction Publishers, 1991), 131.
9. Ernest L. Boyer (foreword), *Campus Life: In Search of Community. A Special Report* (Princeton, NJ: Carnegie Foundation for the Advancement of Teaching, 1990). The report was the result of collaboration with the American Council on Education and the National Association of Student Personnel Administrators.
10. Peter Wood, *Diversity: The Invention of a Concept* (San Francisco, CA: Encounter Books, 2003).
11. Alexander Astin and others, *The Student Learning Imperative: Implications for Student Affairs* (Lexington, KY: American College Personnel Association, 1994), http://www.housing.berkeley.edu/student/ACPA_student_learning_imperative.pdf (accessed February 14, 1996).
12. Alan Charles Kors and Harvey A. Silverglate, *The Shadow University: The Betrayal of Liberty on America's Campuses* (New York: The Free Press, 1998).
13. Bill Rivers, *Never Mind Tocqueville, Let's List Slurs* (Princeton, NJ: The National Association of Scholars, February 26, 2009), http://www.nas.org/articles/Never_Mind_Tocqueville_Lets_List_Slurs (accessed February 27, 2012).
14. I have drawn freely in this account of the University of Delaware program on Adam Kissel's well-documented "Sex, Lies, and Residence Life: Delaware's Thought Reform," *Academic Questions* 22, no. 2 (Spring 2009): 181–99.
15. Keith E. Edwards and Kathleen G. Kerr, "Sustainable Development: Towards Healthy Environments, Economic Strength, and Social Justice," *Tools for Social Justice Conference*, Kansas City, Missouri, November 13, 2006, http://74.125.47.132/search?q=cache:oTqBQmsn_28J:www.myacpa.org/task-force/sustainability/docs/TFSJ%2520Sustainability.ppt+sustainability+%22Kathleen+Kerr%22&cd=1&hl=en&ct=clnk&gl=us (accessed February 27, 2012).
16. Anonymous, "Rodney Complex Curriculum Executive Summary, 2007–2008," July 2007. This unpublished document has a cryptic attribution, "Developed by Sendy E. Guerrier, Rodney Complex Coordinator; Adjusted

by Licinia B. Kaliher, Rodney Complex Coordinator." Nonetheless, some of the text is clearly by Kathleen Kerr. The Foundation for Individual Right in Education posted a PDF copy of the document here: http://www.thefire.org/public/pdfs/81f9cbba277e955afa9823f0ebdce026.pdf (accessed February 27, 2012). The Rodney Complex is a group of dormitories at the University of Delaware.

17. Edwards and Kerr, "Sustainable Development."
18. Todd Cellini and others, *Toward a Sustainable Future: The Role of Students Affairs in Creating Healthy Environments, Social Justice, and Strong Economies* (Washington, DC: ACPA, 2008).
19. Tom Wood, *The Communitarian ResLife Movement* (Princeton, NJ: The National Association of Scholars, July 18, 2008) http://www.nas.org/articles/The_Communitarian_Res_Life_Movement (accessed February 27, 2012).
20. Too many to list here but of particular note, John Blewitt and Cedric Cullingford, eds., *The Sustainability Curriculum: The Challenge for Higher Education* (London: Earthscan, 2004). Blewitt's and Cullingford's collection suggests that the sustainability movement was further along in gaining a place in British universities than it was in the United States at that time. Much of the rhetoric and conceptual apparatus of the movement is the same, but Blewitt is especially interested in "lifelong learning." Another notable book is Ann Rappaort and Sarah Hammond Creighton, *Degrees that Matter: Climate Change and the University* (Cambridge: MIT Press, 2007). Rappaort and Creighton are mostly concerned about how to cut energy consumption on campus. They thus represent a side of the sustainability movement that I do not give much attention in this paper: ingenuity directed at the practical task of energy conservation. Rappaort and Creighton take for granted that climate change is real and that we need to get on with the work of making the campus "green."
21. David W. Orr, "Foreword," xiii; Andres R. Edwards, *The Sustainability Revolution: Portrait of a Paradigm Shift* (Gabriola Island, British Columbia, Canada: New Society Publishers, 2005).
22. David W. Orr, *Earth in Mind: On Education, Environment, and the Human Prospect* (Washington, DC: Island Press, Tenth Anniversary Edition, 2004), quotations from Introduction to new edition, xii, xiv; "What Is Education For?" 9–12; "The Problem of Education," 26 and 33.
23. Peter Wood, *Enchanting Sustainability* (Princeton, NJ: The National Association of Scholars, January 26, 2009), http://www.nas.org/articles/Enchanting_Sustainability (accessed February 27, 2012).
24. Peggy Barlett, "Reason and Enchantment in Cultural Change: Sustainability in Higher Education," *Current Anthropology* 49, no. 6: 1077–98.
25. http://www.aashe.org/highlights/press-releases/digest08 (accessed February 27, 2012).
26. *American College and University Presidents Climate Commitment* (Boston, MA: Second Nature, 2009), http://www.presidentsclimatecommitment.org/html/commitment.php (accessed February 27, 2012).
27. Craig Idso and Fred Singer (lead authors), *Climate Change Reconsidered: 2009 Report of the Nongovernmental International Panel on Climate Change* (Chicago, IL: Heartland Institute, 2009).

28. Ashley Thorne, "Proven Commitment to the Climate," The National Association of Scholars, April 9, 2009, http://www.nas.org/articles/Proven_Commitment_to_the_Climate (accessed February 27, 2012).
29. Niles Barnes, moderator, *Climate Action Planning Wiki* (Lexington, KY: Association for the Advancement of Sustainability in Higher Education, 2009), http://www.aashe.org/wiki/climate-planning-guide/introduction.php (accessed February 27, 2012).
30. The term "Second Nature" was a favorite of Bookchin's. I am not clear whether the organization's founders knew they were signaling the anarchist origins of their movement.
31. Joseph Ben-David, *The Scientist's Role in Society: A Comparative Study* (Englewood Cliffs, NJ: Prentice-Hall, 1971), 184–85.

11

The University and Public Education

Michael Ben-Chaim

Can Education Be Standardized?

In modern liberal democracies, education is widely recognized as a basic human right. Government must respect this right and allow citizens to obtain the education they consider appropriate for themselves, and especially their young. As government becomes more involved in the provision of education, however, teaching and learning are more integrated into a mandatory institutional framework. What happens when the right to education and the obligation to conform intertwine?

It is a common saying that, while it is possible to pull a horse to a bucket of water, it is impossible to force it to drink from it. Compulsory education in various degrees and forms is, nevertheless, a common phenomenon in the civilized world. One of the principal ways of its administration is standards of education that specify what students should know and assess what students' acquired knowledge actually is. Standardized education, then, hinges on the availability of practically unambiguous, impersonal, and discrete articles of knowledge. Furthermore, in consideration of fairness and equity, this knowledge must be impartial and indifferent to the cultural sensibilities and values that distinguish individuals, families, and communities from one another.

Is such knowledge really attainable? According to the conventional wisdom of modern educators, it is in fact generated by modern scientific research. It can and should therefore be transmitted to schools and taught in the classroom by competent teachers. Since the rise of progressivism in late nineteenth century, natural science has been widely regarded as a model of human learning. The sciences as a whole have become the principal source of theories and methods in education. However, while American science assumed worldwide leadership during the twentieth century, American public education in general, and especially school mathematics and science, has remained entrenched in mediocrity. Furthermore, the number of students with demonstrated average or above average intelligence who nevertheless fail to meet educational standards and receive, as a result, special educational assistance has steadily grown. The cost of college remediation and preparedness of high school graduates has risen accordingly. Following the 1983 National Commission on Excellence in Education publication, *A Nation at Risk*, a growing number of educators and public officials have warned that the deep-rooted mediocrity in elementary and secondary education will eventually jeopardize the advancement of American science itself.[1]

The discrepancy between the soaring achievements of science and the sore state of school education is especially puzzling in light of the unrelenting capacity of the former to generate innovation in the manufacturing industries, healthcare, and information technology. How is it that science has become an engine of growth in these areas, and at the same time has failed to inspire excellence in the classroom? Given the prevalence of the paradigm of knowledge-based standards of education, endeavors to find remedy for the chronic ills of public education have been commonly predicated on the assumption that, since science had done its part and delivered the knowledge, the fault must lie with its distributors at the school level, especially teachers, their training, and the administration of their work.

The present paper offers a different perspective on the problem. I argue that schools could perform considerably better if they were not bound by standards of impersonal, impartial, and itemized knowledge. Such knowledge does not exist. Scientific work at its best is and always was a personal, passionate, value-driven, reflective, and imaginative intellectual endeavor, academically interdisciplinary and in tune with its social and cultural environment. The requirement to teach and learn standardized knowledge conveys a misrepresentation of science as a model of human learning and is incompatible with education. As will be shown in the rest of this paper, packaging scientific research as standardized knowledge is an artifact of ideology rather than the outcome of educational research or policy. The seedbed of this ideology is higher education, and, more specifically, the modern university, rather than the teachers' seminary and classroom that, by and large, have simply followed suit.

The University's Twofold Role in Society

"A university," writes Eric Ashby, is an institution that "preserves, transmits, and enriches learning."[2] The university, in other words, is an agent of education. Throughout its history, it was always the largest and most prominent institution in Western civilization dedicated to the advancement of formal and literacy-based education. However, the university's role in society was never exclusively confined to the advancement of learning. The university traditionally affirmed the political order within which it operated, and had to combine its educational endeavors with the mission to legitimate political power.

Rising in the twelfth century in response to the expansion of European towns and trade, the growing need for a professional cadre in civil and ecclesiastic administration, and the encounters between Christian and Muslim civilizations, European universities were islands of literate culture surrounded by an ocean of illiterate, agrarian communities. These islands were the seat of a class of literate specialists. The difference between their academic culture and that of the surrounding illiterate communities of peasants and artisans was profound, especially because it was entrenched by political and economic divisions that rigidly separated the ruling class from the rest of society.

The centralization of literate culture in the university, and the centralization of political power in state or church government, did not always go hand in hand. Tensions between gown and town were common, and universities were often entangled in rivalries among local rulers. In general, however, the university was always dependent on the support and protection of governing elites. In return, it provided members of these elites the professional and doctrinal means to establish the legitimacy of their status and the authority of their rule. Thus, the forms of literacy and literate learning that were transmitted and enriched within the university were also intended to secure conformity and submission beyond its walls. Traditionally, then, the social role of the university was twofold, namely, to educate its students and to articulate and propagate the ideology of the governing forces in society.

Why Education Is Incompatible with Ideology

Education and ideology are not always easily distinguishable. Both aim to mold the human capacity to distinguish between right and wrong, and thereby to guide thought and behavior. The educator and ideologue often hold on dearly to their respective ideas, even and perhaps especially, when contested by others.

Considered from a sociological viewpoint, however, education and ideology have distinctively incompatible functions. Education stems from the disposition of adults, and especially parents, to share their experience, modes of thought, and ways of life with the younger generation, and the complementary disposition of the latter to accept this form of assistance. In this respect, education can be seen as a cultural derivative of the biological phenomenon of the successful adaptation of reproductive mechanisms. In another respect, however, education is a uniquely human characteristic that testifies to the overwhelming dependence of the less experienced on the more experienced members of the species, and the innate human capacity to communicate and share experience to increase the likelihood of survival. The meager evidence for the survival of feral children, and the extreme physiological and behavioral depravity of the few who were identified in the past, amply demonstrate that human beings must learn from other human beings how to become human beings. Education in all its forms is the institutional mechanism that is set to satisfy this condition.

Education enhances solidarity among people, especially between the older and younger members of community. The need for ideology, by contrast, arises when two or more belief systems are perceived to be in competition for hegemony and are, therefore, in conflict. Originating in the political arena, the function of ideology is to enable one group of people to secure its control over another group despite actual or potential conflicts of interest that separate them. For example, the medieval ideology of divine right of kings provided rulers with a superhuman authority without which their power was more likely to be challenged by popular discontent. As illustrated by this example, ideology provides ultimate validity and justification for particular modes of thought and action of a politically dominant group. Ideology is, accordingly, codified in a universal and impersonal idiom that is explicitly uninfluenced by, and independent of, the distinctive qualities, needs, or aspirations of individual persons and their special life circumstances.

Education empowers human agents by enabling them to benefit from the experience of others. Since instructor and student are likely to differ from one another in some ways that are relevant to their collaborative endeavor, it is in their common interest to comprehend how the experience that proved useful to the one under certain circumstances may still be useful to the other under somewhat different circumstances. Hence, awareness of and respect for diversity are inherent to the educational process. Furthermore, however methodical and meticulous instruction is, mastering a new experience is ultimately the student's responsibility and achievement. The instructor simply cannot do it on behalf of the student. Education inherently involves the student's reflective thinking and self-mastery, and is therefore an exercise in the cultivation of personal autonomy.

In contrast with education, the successful ideology diminishes personal autonomy and diversity. In its endeavor to monopolize validity and justification, ideology requires individuals to assent to a particular view of the world and their place in it. Hence, irrespective of the language it uses to characterize its redeeming value, ideology inevitably diminishes the capacity of individuals, and deprives them of the right to figure out for themselves what their life experiences mean to them, what their goals ought to be in light of their experience, and how they ought to achieve their goals. Experiences and decisions that fail to converge with the dictates of ideology appear, to the converted, as irrational and absurd. Personal autonomy empowers the individual as a reflective agent. Ideology renders reflective thinking redundant, since it claims to monopolize the key to identify the right questions one might ask and the method to arrive at the respective answers.

In order to attract potential believers, ideology wraps itself in the promise of widely accepted ideals. This is its self-proclaimed redeeming value. It concomitantly portrays the era prior to its rise as a provisional and illusory bygone age which the believer had to renounce in order to find the light of truth. To undermine the credibility of rival belief systems, ideology claims that any other way to pursue the ideals it upholds is bound to lead the innocent follower astray. The promise of salvation coupled with the curse of damnation equips ideology, as Gellner noted, with "both a trap and a bait: an entry-inducing and an exit-hampering mechanism."[3]

From Faith through Secularism to Knowledge

Until the eighteenth century, the dominant ideologies that were propagated by the university reflected in one way or another the cultural supremacy of the Christian religion in Western civilization. Religious diversity and respect for personal autonomy concerning matters of faith were tolerated only within limits set by ecclesiastical and civil authorities. The fundamental tenets of Christianity were presumed to be the object of universal conformity.

During the nineteenth century, universities gradually adopted the view that religion, in general, conveyed at its core beliefs and opinions that could not but be construed as expressions of personal choice and changing social norms and customs. The academic discourse on issues of faith was pursued within a secular cultural framework. Secular culture was by no means necessarily antireligious. Rather, it was based on the assumption that it was possible and desirable to engage in the study of the universe and humanity's place in it without the presuppositions of any particular belief system. With the decline of the authority of the clergy, the expansion of national bureaucracies, and the industrialization of the economy, university education was meant to lead to greater secular knowledge rather than a more enlightened faith in God.

Within the secular framework that was advocated by academic discourse, the transition from belief to knowledge reflected the strategic reorganization of the university into an institution of scientific education that unified research and teaching. From the middle of the nineteenth century, the combined effort of scientific research and instruction in the single vocation of the university professor has been the most important theme in university reforms and in the establishment of higher education the world over. From a pedagogical viewpoint, this transition conveyed the conviction that teaching by innovating researchers would be more inspiring for students than the traditionally systematic instruction by competent teachers, even if it did not provide full coverage of any particular academic field.[4]

The reorganization of the university implied, accordingly, that belief and knowledge were not simply two equally legitimate or valuable modes of thought. Knowledge was superior to belief. Moreover, the superiority of knowledge could not simply be conceived as a matter of personal choice or opinion. That would be counterproductive, since it would merely reinstate the sovereignty of belief. Could the superiority of knowledge be theorized on the basis of universal and impartial principles?

The search for a theory of knowledge, explained Bertrand Russell, is motivated by "the hope of being able to distinguish trustworthy beliefs from such as are untrustworthy . . . At first sight it might be thought that knowledge might be defined as belief which is in agreement with the facts. The trouble is that no one *knows* what a belief is, no one *knows* what a fact is, and no one *knows* what sort of agreement between them would make a belief true."[5] In the context of everyday life—at the university or outside it—the difference between more and less trustworthy beliefs is always a matter of personal judgment, custom, and circumstances. As Russell suggested, the search for a criterion that would demarcate knowledge from belief is inevitably regressive, for, if we do not know what knowledge is, we will never be able to know that we have found it. From an educational viewpoint, then, the quest for a criterion that would demarcate knowledge from belief appears futile. Nevertheless, versions of the holy grail of knowledge were embraced with enthusiasm by university professors seeking to defend the authority of academic teaching at the age of science.

Ideologies of Knowledge in the Twentieth Century

The most influential ideologies of knowledge in the twentieth century were logical positivism and scientific realism. Evolving from the writings of early modern experimental philosophers, logical positivism reinvigorated Enlightenment images of the progress of humanity beyond ages of superstitious thought.[6] The positivist portrayal of a radical break from the superstitions of the past suggested that science was not bound by tradition in its search for knowledge. It could transcend the educational process of the transmission and enrichment of learning from generation to generation. According to positivism, learning from experience sufficed, and that was the motto and method of modern science. Scientific progress exemplifies the power of the human intellect disciplined by rational inference and the rigorous assessment of empirical evidence.

The scientific method, in its logical positivist formulation, demarcates knowledge from other expressions of human thought and culture. The distinction hinges on the question of meaning. According to positivism, reason and experience render human thought meaningful. Technically, the meaning of a sentence is its method of verification. To become meaningful, a sentence must be either logically valid or empirically testable. It follows from this verification criterion that religious, metaphysical, moral, and esthetic ideas are cognitively insignificant because they transgress the realms of logic and sensory data.

An obvious objection to the positivist definition of the scientific method, as Hilary Putnam suggested, is to point out that the criterion of meaning upon which it rested is self-refuting: The criterion itself is a normative claim that prescribes a specific view on meaning and knowledge, and cannot be verified by experience. Since it is neither a logical inference nor an empirically testable proposition, it is cognitively insignificant.[7] What rendered the positivist method of science valuable, despite its incoherent foundations, was its ideological function. The bait it offered could not be easily refused: Acknowledge the authority of scientific knowledge, or remain in ignorance and blinded by assertions that are meaningless because they are nonscientific. The offer comes with an invitation to embrace science as a dynamic system of empirical facts. The system continuously expands to include more and more parts of the observable universe, and is rendered more efficient by the continuous formulation of abstract theoretical correlations of factual findings. Its progress neither depends on any particular cultural values and ideals, nor requires their approval, and is therefore the ideal object of universal assent. Scientific knowledge is value-neutral. In this specific respect, it is objective, and that is precisely its value.

A trap, nevertheless, accompanies the bait. Value-neutral science, as Max Weber observed, "is meaningless because it gives no answer to our question, the only question important to us: 'What shall we do and how shall we live?'"[8] If the study of science is a study of indifferent facts, the student may respond to the invitation with indifference. If it is impersonal, the student may not integrate it with her own intellectual development. If it is dissociated from the lives of individuals and communities, then, even if they accept its validity, they may not consider it seriously as a desirable component in their education.

Though positivism still echoes in the contemporary classroom, its academic reputation has declined. One of its grave weaknesses is its notion of scientific theory as merely a convenient correlation between empirical findings. Scientifically literate people believe in the scientific theories they are familiar with, and in this respect they do not differ from others who believe that the world in which they live is more than what meets the naked eye. Scientific realism caters specifically to believers, and divides them into two mutually exclusive groups: Those who believe that scientific theories are truthful, or at least approximately truthful depictions of physical reality, on the one hand, and those who hold on to beliefs that fail to cohere with the findings of science, on the other. Following the footsteps of positivists, scientific realists uphold the binary opposition between knowledge and ignorance in their evaluation of the fundamental difference between the two groups of believers. Those who believe in science are the enlightened; the others are cognitively doomed.

More specifically, scientific realism claims that the product of successful scientific research is by and large knowledge of a theory-independent reality. The student who examines a scientific textbook ought, accordingly, to consider the claims it contains about, for example, atoms and their subatomic particles, molecules and their chemical structure, or the mechanism of chemical reactions, as corresponding to the states and properties of real objects. Moreover, the belief that true information is provided by the textbook is justified even in those cases in which the relevant aspects of reality are not directly observable, the research protocols and instruments that are used to obtain the information are fallible, or the empirical findings of the reported research are approximate.[9]

The fallacy of scientific realism, which in its general form is conveyed by the aforementioned quotation from Russell, is that it does not provide the means for examining whether or not a particular theory is indeed true. If we knew what the physical reality was, we could have determined how well it is described by a particular theory. But then we would not have needed the theory in the first place. Scientific research is a valuable intellectual enterprise precisely because we do not know what the physical reality is. Hence, accepting scientific realism and conforming to the dictum that science approximates the truth does not help us understand better either science or physical reality.

Scientific realism is, nevertheless, ideologically effective. The ultimate justification for the belief in the success of science, according to scientific realism, has nothing to do with the believer, and has everything to do with the physical reality. The justification derives its credibility in part from the following analogy: If you find a penny in your pocket, the best explanation for your finding is that the coin was really there in the first place! The justification concomitantly explains why other belief systems that do not accord with science are not successful and ought to be discarded.

The ideological trap of scientific realism lies in portraying the success of science in terms of its alleged relation to reality rather than its characteristics as a human, intellectual endeavor. Learning science implies, then, turning a blind eye to the personal, social, and cultural meanings of success and failure in scientific research. For example, when we learn biology, we learn about the reality of life processes rather than from the intellectual pursuits of biologists. Accepting the truth about this reality secures conformity to the science of biology. It suggests, moreover, that what scientists do does not really matter; the important thing is that the results of their endeavors happen to approximate the truth.

Education without Ideology

Ideologies of knowledge such as positivism and scientific realism espouse an image of human learning that is incompatible with education. Constituting the basic mechanism in the evolution of humanity, education, as I suggested above, is an interpersonal experience of learning from human beings how to become a human being. The ideological image, by contrast, undermines the import of learning from others and claims that learning is a process of interaction between the human subject aided by innate reason and sense organs, on the one hand, and the objective world, on the other. According to positivism, this is the most, and perhaps the only rational mode of gaining knowledge; according to scientific realism, it is the most effective, and perhaps the only way of getting to know the truth about the physical world.

The ideological image of learning has also exercised considerable influence in specific fields of academic scholarship, notably cognitive psychology. For example, Jean Piaget, whose writings have exerted considerable influence on modern child psychology and education, proposed that adaptation to the physical environment was the principal, if not the only, mechanism of the early growth of intelligence. He suggested, more specifically, that the intelligence of children matures through a developmental course of several stages of adaptation that begins with sensorimotor behavior and culminates with the capacity to abstract and draw inferences in a systematic manner. In none of these stages of development, the child's learning from other human beings plays any significant role.

According to Piaget, children of seven to eleven years old are likely to rediscover on their own the principle of the conservation of matter, will independently acquire hypothetical-deductive reasoning in their early teens, and, if given the adequate opportunity, could rediscover on their own the basic principles of Newtonian mechanics as they got older.[10] Since Piaget observed children who were in regular contact with adults, he could not properly test his theories. Moreover, in contrast with Piaget's hypothetical young learners, Isaac Newton, who was by no means generous when it came to acknowledging his debt to other scholars, disclosed to his colleague and rival at the Royal Society of London, Robert Hooke, that, "if I have seen further, it is by standing on ye sholders of giants."[11] Given the impact of modern ideologies of knowledge on the public imagination, however, it is not surprising that the popular image of Newton is the lonely spectator of an apple tree, rather than the dedicated student from the University of Cambridge who took notes meticulously while reading the works of René Descartes or Robert Boyle.

Human agents and scientists in particular undoubtedly learn by means of observation and experiment; they do it effectively, however, because they stand on the shoulders of giants. Newton's metaphorical account, then, is a reminder that empirical research need not be construed ideologically as a special method that leads to superior knowledge *because* it is devoid of personal, social, and cultural import. Starting with this metaphor and following the works of Adam Smith and, more recently, Michael Polanyi and Joseph Ben-David,[12] I propose an evolutionary model that is based on three principal assumptions that show how learning from others and from experience are two complementary aspects of a unitary educational process.

The first assumption of the model is cultural differentiation and diversity. At any given time, individual agents engage in problem-solving and other task-oriented activities that require more or less special resources and tools. The activities commonly involve applying the tools to the transformation of the resources into valuable products, or goods. Human culture is the totality of all these diverse transformations.

The second assumption is that every transformation that is carried out by a particular person is a model of learning and a learning opportunity for another. Hence, learning is necessarily a selective process. Instructor and student are likely to select those transformations that were proved beneficial to some people in the past, and therefore might be beneficial to others in the future. Hence, the principal object of selection is values—values that are embodied in information, skills, techniques, methods, and products. The selection triggers the transmission of learning from one person to another, in other words, its cultural replication.

The third assumption of the model is that replication necessarily involves personal autonomy. The student who selects the activity may seek to replicate its record of success as best as she can, but no one else can do that on her behalf. The student may benefit from the assistance of an instructor or a mentor, but must figure out herself how to integrate the activity she aims to replicate with her own habits of thinking, skills, beliefs, and background knowledge of the situation. Moreover, the exercise of autonomy is transformative, empowering the individual learner and at the same time modifying the transmitted activity.

According to this model, then, intellectual diversity and autonomy are the engines of the growth of learning. The outcome of this growth is an expanding web of multiple and diverse educational opportunities that are transmitted and enriched by individual agents interacting with one another and with their environment. Newton's contributions to physical sciences are eminent examples of learning enriched by transmission. The application of mathematical tools to the explanation of physical phenomena, which was invigorated by scholars since the sixteenth century and in which Newton later took part, is another example on a larger scale, comparable perhaps to the transformation of twentieth-century biology that was effected by the application of physical chemistry to the study of heredity. Indeed, the history of science and technology is replete with examples of major and minor innovations that were the result of the crossover of modes of explanation, calculation, and representation, as well as of manufacturing machines, laboratory apparatuses, and instruments of observation and measurement.[13]

Innovations in science and other cultural traditions are models of education rather than conformity to standards set by ideology. Can the lessons these models teach be extended to the education of all children? As long as the university operated in a society that was by and large illiterate, it was practically impossible to replicate models of learning that were transmitted and enriched within the university outside its walls. In those days, the vast majority of children received their education in informal settings that were established by their extended families and communities. Local lore and customs were transmitted and enriched from generation to generation in ways that were not significantly affected by the university. For this reason, the ideologies that the university propagated could not be effectively integrated with local education and therefore could not seriously undermine its quality.

Paradoxically, the ideologies of knowledge that have been propagated by the university since the nineteenth century were more dominant in society than earlier ideologies because, during this period, the levels of literacy in society increased rapidly. Conformity to standardized knowledge could be officially imposed in every town and village that was integrated into the system of public and literate education. Scientific research could have provided a model of intellectual growth and development for the wider population of students, but instead it has been systematically used as a license to decree what they "should know." Thus, the diffusion of literacy in society has been carried out by ideological means that purported to increase education but in effect have disabled the cultivation of students' autonomy and diversity.

Following the continuous expansion of literacy in society throughout the twentieth century, a seamless educational path that begins in preschool and continues through graduate school is today a goal for a large and growing segment of the population. Students who take this path should not be derailed by ideologies and institutions that encourage conformity at the expense of intellectual diversity and autonomy. Educators stationed along this path should help students understand how people learn rather than propagate a standardized package of impersonal knowledge. The university of the twenty-first century should take a leading role in the collective endeavor to achieve these objectives.

Notes

1. United States National Commission on Excellence in Education, *A Nation at Risk: The Imperative for Educational Reform* (Washington, DC: The Commission, 1983); David T. Gordon, ed., *A Nation at Risk? American Education 20 Years after a Nation at Risk* (Cambridge, MA: Harvard University Press, 2003).
2. E. Ashby, "Ivory Towers in Tomorrow's World," *Journal of Higher Education* 38 (1967): 417.
3. E. Gellner, "Notes towards a Theory of Ideology," in *Spectacles & Predicaments* (Cambridge: Cambridge University Press, 1979), 117–32, Quoted from 121.
4. J. Ben-David, "Science and the University System," *International Review of Education* 18 (1972): 44–60.
5. B. Russell, "Theory of Knowledge," *The Encyclopaedia Britannica* (London/ New York: The Encyclopaedia Britannica Company, 1926).

6. A. J. Ayer, *Logical Positivism* (Glencoe, IL: Free Press, 1959); M. Ben-Chaim, *Experimental Philosophy and the Birth of Empirical Science: Boyle, Locke, Newton* (Aldershot: Ashgate Publishing, 2004).
7. H. Putnam, "Philosophers and Human Understanding," in *Realism and Reason* (Cambridge: Cambridge University Press, 1985), 184–204.
8. M. Weber, "Science as a Vocation," in *From Max Weber: Essays in Sociology*, ed. H. H. Gerth and C. Wright Mills (New York: Oxford University Press, 1958), 143.
9. P. Churchland and Clifford A. Hooker, eds., *Images of Science: Essays on Realism and Empiricism, with a Reply from Bas C. van Fraassen* (Cambridge: Cambridge University Press, 1980).
10. J. Piaget, *The Child's Construction of Reality* (London: Routledge and Kegan Paul, 1955); J. Piaget and R. Garcia, *Psychogenesis and the History of Science* (New York: Columbia University Press, 1989).
11. I. Newton, *The Correspondence of Isaac Newton*, ed. H. W. Turnbull, J. F. Scott, A. R. Hall, and L. Tilling, 7 vols. (Cambridge: Cambridge University Press, 1959–1977), I, 416–17, February 5, 1675/1676.
12. A. Smith, *Lectures on Jurisprudence*, ed. R. L. Meek, D. D. Raphael, and P. G. Stein (Oxford: Clarendon Press, 1978), 347–48; *Inquiry into the Nature and Causes of the Wealth of Nations* (Harmondsworth: Penguin Boos, 1986), 115; M. Polanyi, *Personal Knowledge: Towards a Post-Critical Philosophy* (New York: Harper Torch books, 1964); M. Polanyi and H. Prosch, *Meaning* (Chicago, IL: The University of Chicago Press, 1975); J. Ben-David, "Roles and Innovations in Medicine," *American Journal of Sociology* 65 (1960): 557–69; "Social Factors in the Origin of a New Science: The Case of Psychology," *American Sociological Review* 31 (1966): 451–65.
13. J. S. Fruton, *Molecules and Life. Historical Essays on the Interplay of Chemistry and Biology* (New York: Wiley-Interscience, 1972); M. J. S. Rudwick, *The Great Devonian Controversy: The Shaping of Scientific Knowledge among Gentlemanly Specialists* (Chicago, IL: The University of Chicago Press, 1985); G. Basalla, *The Evolution of Technology* (Cambridge: Cambridge University Press, 1988); M. Ben-Chaim, "The Empiric Experience and the Practice of Autonomy," *Studies in History and Philosophy of Science* 23 (1992): 533–55; "Empowering Lay Belief: Robert Boyle & the Moral Economy of Experiment," *Science in Context* 15 (2002): 51–77; T. Lenoir, *Instituting Science. The Cultural Production of Scientific disciplines* (Stanford, CA: Stanford University Press, 1997); J. Ziman, "Evolutionary Models for Technological Change," in *Technological Innovation as an Evolutionary Process*, ed. Ziman (Cambridge: Cambridge University Press, 2000), 3–12.

12

In Lieu of the Epilogue: Joseph Ben-David in Photographs

Miriam Ben-David

A note from the editor: In her presentation, Miriam Ben-David called her late husband "Yossi." This was the name his friends used. After I finished my PhD, he told me to call him by this name too, but I never could. He remained for me then, and remains for me still, "Professor Ben-David," a figure of too much authority to allow such familiarity. I therefore substituted "JBD" for every instance of "Yossi" in Miriam's comments to photographs she presented.

It is also worth mentioning that Joseph Ben-David was born Joseph Gross. His father, David Gross, perished in the Holocaust, together with Joseph's mother and two youngest brothers. Thus he changed his name to Ben-David—the son of David. The experience of his family, obviously, affected his views.

Joseph Ben-David (1920–1986)

The Early Stages—Győr

The Gross Family

A short Red Cross note sent to Palestine by his parents

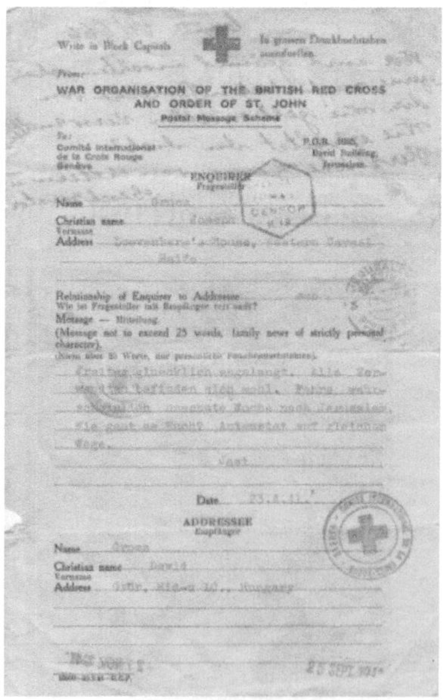

Write in Block Capitals — In grossen Druckbuchstaben ausfuellen.

From:
WAR ORGANISATION OF THE BRITISH RED CROSS AND ORDER OF ST. JOHN
Postal Message Scheme

To: Comité International de la Croix Rouge Genève

P.O.B. [illegible], David Building, Jerusalem.

ENQUIRER
Fragesteller

Name [illegible]
Christian name [illegible]
Vorname
Address [illegible]

Relationship of Enquirer to Addressee [illegible]
Wie ist Fragesteller mit Empfänger verwandt?

Message — Mitteilung
(Message not to exceed 25 words, family news of strictly personal character).

[illegible]

Date [illegible]

ADDRESSEE
Empfänger

Name [illegible]
Christian name [illegible]
Vorname
Address [illegible]

Uri (JBD's youngest son) in front of the Gross family's house in Győr (from family trip 1998)

The Benediction Gymnasium, where JBD got his secular education

The Beginning of His Career

JBD with a group of students in the old city of Jerusalem

JBD began his professional career as a probation officer under the British mandate in Palestine.

The pass to an academic career

Farewell party of the social workers programs for students from the British colonies, at L.S.E.

L.S.E was where Edward Shils introduced him to modern sociology, and when and where JBD's passion to the field began

JBD with Ralf Dahrendorf at the Center for Advanced Studies in the Behavioral Sciences at Stanford (1958)

JBD with colleagues in Japan. From an early age, still as a teenager in Hungary, he was enchanted with the Japanese culture.

With his old-time friend and colleague Jacob Katz at JBD's grandson's Brith (circumcision ritual).

JBD in His Own Environment

Hiking in the Judea mountains

At his home office desk in Jerusalem

JBD and His Family

JBD and Miriam in Chicago

JBD and Family (Berkeley 1964)

JBD and Miriam with four of their grandchildren (1985)

His Personal Legacy

As a person and with his family, JBD somewhat signifies the revival of the Jewish people after the Holocaust.

Both his parents and two young brothers were massacred by the Nazis. His brother Aharon, one of the two brothers that survived Auschwitz, was later killed by a mine during military service in the IDF.

These tragic affairs were part of JBD, yet they have not affected his positive attitude to life nor have they diminished his belief in the future.

Joseph Ben-David 1920—1986

Every man has a name
Given him by God
And given by his father and his mother

Every man has a name
Given him by his stature and his way of smiling,
And given him by his clothes.

Every man has a name
Given him by the mountains
And given him by his walls

Every man has a name
Given him by the planets
And given him by his neighbors.

Every man has a name
Given him by his sins
And given him by his longing

Every man has a name given him
by those who hate him
And given him by his love

Every man has a name
Given him by his holidays
And given him by his handiwork

Every man has a name
Given him by the seasons of the year
And given him by his blindness

Every man has a name
Given him by the sea
And given him
By his death.

Miriam Ben-David concluded her presentation with this poem by Zelda [Mishkovsky].

Contributors

Michael Ben-Chaim earned his PhD in history and philosophy of science from Cambridge University in 1991. His research work has centered on historical sociology of science, with special reference to the early modern period. His book, *Experimental Philosophy and the Rise of Empirical Science*, was published in 2005. He now specializes in learning disabilities, and his work centers on the relationship between learning and reading. He is currently a faculty member of Eagle Hill School, Hardwick, Massachusetts.

Miriam Ben-David was born in 1928 in Budapest, Hungary. During World War II, when the Holocaust reached Hungary under the German occupation, she was hidden by a Christian family. Following that, she fled from Hungary with her parents and arrived in Palestine in the summer of 1944. She finished her schooling in Palestine and studied psychology at the Hebrew University and in the United States. After completing her studies—MA in psychology and education—she worked, first, as an educational psychologist in the Israeli school system, and then as a clinical psychologist and family therapist, as well as supervisor in a mental hospital and mental health center and in private practice in Jerusalem. After retirement from public service, she joined the staff of "AMHA," a clinic for psychosocial services for Holocaust survivors and "second generation." In 1998, Mrs Ben-David set up a project in Hungary, named "Dialogue for Tolerance." The aim of the project has been not only to preserve the memory of the Holocaust as a historical event but also to apply it as a deterrent example of the inhuman consequences of racism, in general, and anti-Semitism, in particular. Begun as an independent voluntary program, it has become affiliated with the Holocaust Museum and Documentation Center of Hungary since 2009. Miriam had been married to Joseph Ben-David since 1947 till his death in 1986.

Mario Coccia is an economist at the National Research Council of Italy visiting scholar at Georgia Institute of Technology, and a visiting professor of industrial organization at the *Università degli Studi del Piemonte Orientale.* His research focuses on technometrics and on the technological impact of innovation and the driving forces of structural and economic change. He is the author of numerous articles.

Brendan Dooley (Chicago, PhD, 1986) is currently professor of Renaissance studies at University College Cork, Ireland. Previous appointments have included Harvard University and Jacobs University Bremen. He has held fellowships at the Institute for Advanced Study, American Academy in Rome, and European University Institute. He is particularly interested in the history of knowledge, including the problems of production, diffusion, and storage. His publications include *The Social History of Skepticism: Experience and Doubt in Early Modern Culture* (The Johns Hopkins University Press, 1999), *Science and the Marketplace in Early Modern Italy* (Lexington Books, 2001), as well as *Morandi's Last Prophecy and the End of Renaissance Politics* (Princeton University Press, 2002). He is currently working on *Brill's Handbook on Renaissance Astrology.*

Yaron Ezrahi is Gersten Family Professor of Political Science at the Hebrew University of Jerusalem. His research interests cover comparative politics and democracies; modern political philosophy; science, technology, and politics; cultural policies; and culture and democracy. He is the author of *The Descent of Icarus: Science and the Transformation of Contemporary Democracy* (Harvard University Press, 1990), *Rubber Bullets: Power and Conscience in Modern Israel* (University of California Press, 1998), and *Israel: Towards a Constitutional Democracy* (with Mordechai Kremnitzer, 2001, in Hebrew).

Claudius Gellert has held academic appointments at Cambridge, Florence, Halle-Wittenberg, Harvard, Humboldt, Klagenfurt, Ludwig-Maximilians-Universität München, and Reading Universities. He has been involved with many educational and policy organizations, including the German Academic Exchange Service, the OECD, the *Wissenschaftsrat,* the BMBF, the European Union's Commission on Higher Education Systems, and the World Bank. He is the author of *Higher Education in Europe* (Jessica Kingsley, 1993), *Diversification of European Systems of Higher Education* (Peter Lang, 1995), and *Innovation and Adaptation in Higher Education* (Jessica Kingsley, 1999).

Liah Greenfeld is university professor and professor of sociology, anthropology, and political science at Boston University, distinguished adjunct professor at Lingnan University in Hong Kong, and the director of the Institute for the Advancement of the Social Sciences. Her books include, among others, *Different Worlds: A Study in the Sociology of Taste, Choice, and Success in Art* (based on her PhD thesis, written under the direction of Joseph Ben-David, Cambridge University Press, 1989), *Nationalism: Five Roads to Modernity* (Harvard University Press, 1992), *The Spirit of Capitalism: Nationalism and Economic Growth* (Harvard University Press, 2001), and *Nationalism and the Mind: Essays on Modern Culture* (Oxford: Oneworld, 2006).

Dmitri Panchenko has been teaching at the Smolny College of Liberal Arts and Science (St. Petersburg State University) and is currently a research fellow at the Helsinki Collegium for Advanced Studies. He is also a member of the editorial board of the journal *Hyperboreus*. His research interests concern Greek intellectual history. He is the author, among other publications, of *Платон и Атлантида.* (Plato and Atlantis) Л.: Наука (1990) and *Thales and the Origins of Theoretical Reasoning* (2005, in Greek).

Nathalie Richard is professor of modern history at the *UNAM, Université du Maine* (Centre de recherches historiques de l'Ouest, CNRS UMR 6258, Le Mans, France). She is the author of *Inventer la préhistoire: Une histoire des débuts de l'archéologie préhistorique en France* (Vuibert 2008) and *Hippolyte Taine. Histoire, psychologie, littérature* (Garnier, forthcoming). She has published, together with Jacqueline Carroy et al., *Alfred Maury, érudit et rêveur. Les Sciences de l'homme au milieu du 19e siècle* (Presses universitaires de Rennes 2007), *Les femmes dans les sciences de l'homme* (Séli Arslan 2005) and *La découverte et ses récits en sciences humaines* (L'Harmattan 1998).

Chikako Takeishi teaches sociology at Chuo University, Tokyo. She did her doctoral studies at Harvard, with the dissertation titled *Rethinking Japanese National Identity: Narratives of Japanese Intellectuals* (1999) and has published on issues related to the theory of nationalism, social relations, and Japanese national identity.

Peter Wood is president of the National Association of Scholars. He is the author of *A Bee in the Mouth: Anger in America Now* (Encounter Books, 2007) and of *Diversity: The Invention of a Concept*

(Encounter Books, 2003), which won the Caldwell Award for Leadership in Higher Education from the John Locke Foundation. He is a graduate of Haverford College, Rutgers University, and the University of Rochester, from which he received a PhD in anthropology in 1987. He previously served as provost of The King's College in New York City, and as the president's chief of staff at Boston University, where he was also a tenured member of the anthropology department.

Index

For Product Safety Concerns and Information please contact our
EU representative GPSR@taylorandfrancis.com Taylor & Francis
Verlag GmbH, Kaufingerstraße 24, 80331 München, Germany